Marie Kelly was born in County Durham; after marrying Bernard, they spent many years in South Africa where he pursued a show business career. On returning to England, Marie became the managing director of an employment agency, eventually leaving to operate her own agency. In 1989, they forsook their London lifestyle for a farmhouse in north Cumbria, where they continue to live with their dogs and other assorted animals. Marie now runs a cottage craft industry from their home.

Breeze
Waif of the Wild

Marie Kelly

HEADLINE

Copyright © 1996 Marie Kelly

The right of Marie Kelly to be identified as the Author of
the Work has been asserted by her in accordance with the
Copyright, Designs and Patents Act 1988.

First published in 1995
by HEADLINE BOOK PUBLISHING

First published in paperback in 1996
by HEADLINE BOOK PUBLISHING

10 9 8 7 6 5 4 3 2 1

All rights reserved. No part of this publication may be
reproduced, stored in a retrieval system, or transmitted,
in any form or by any means without the prior written
permission of the publisher, nor be otherwise circulated
in any form of binding or cover other than that in which
it is published and without a similar condition being
imposed on the subsequent purchaser.

ISBN 0 7472 5264 5

Printed and bound in Great Britain by
Cox & Wyman Ltd, Reading, Berks

HEADLINE BOOK PUBLISHING
A division of Hodder Headline PLC
338 Euston Road
London NW1 3BH

To Joy

ACKNOWLEDGEMENTS

With gratitude I thank warmly Sue Ward, who sacrificed endless lunch hours patiently typing; Ron Nicholson for his photocopying services; and Margaret and Jack Sisson for their constant encouragement and being friends indeed.

Contents

1	Coldside	1
2	The Orphan	3
3	Pleased To Meet You	11
4	Fur, Feathers and Foundling	17
5	The Survivor	29
6	Facts, Fears and Fatigue	37
7	Breeze's Graduation	45
8	Laying the Foundations	51
9	Flora and Fauna	63
10	The Sweet Smell of Success	75
11	Wet, Wet, Wet	83
12	Growing Pains	93
13	'Scours'	105
14	Helping Hands	121
15	Narrow Escape	135
16	'Twiddly Twonks'	145
17	The Yearling	159
18	In the Family Way	175
19	Marking Time	185
20	Dawn and Dusk	193
21	Paradise Lost?	203
22	Spring 1995	213

1
Coldside

Fate brought us to Coldside eventually. It was the forest abutting the farmhouse with which I instantly fell in love but I was unaware then of the riches it would yield. Neither did I anticipate that this woodland was to become my heartbeat.

The need to migrate from the pressures of our London lifestyle wasn't one of those decisions that we considered for too long. It was more like waking up with a violent toothache and announcing, 'I'm off to the dentist, this tooth and I are parting company, today'!

I was running my own successful employment agency, and my husband, Bernard, was in the snakes and ladders business of pop-group management. His daily migraines and clutch of non-organically grown ulcers seemed to be the price he was paying for fast-lane living and eating the occasional plate of ego food. A healthy bank balance being no compensation for an unhealthy body, we determined to get our priorities in order.

'Why give up all you have here?' 'What will you do for a living?' 'Where are you moving to?' Some questions can be answered, others can't, and in respect of all three we could only reply truthfully, 'We haven't a clue.'

BREEZE: WAIF OF THE WILD

With our business ends tied up, house sold to the first viewer and furniture in storage, we packed our suitcases, shoe-horned our four massive canines into the car, and departed for a battery-charging sojourn in Eire. We found it to be beautiful, the people friendly and the climate very wet! Perhaps it was too much of a culture shock to move from the hustle and bustle of London to the absolute quiet of the Dingle Peninsula. Eire seemed right, but it wasn't *quite* right: not the place in which to settle permanently, despite Bernard's Irish origins. Some things are meant to be, and nothing we do will change them.

The estate agents I contacted in Britain soon began responding to my requests for details on a detached property anywhere, with a little land and no near neighbours. Was the Trade Description Act specifically passed for estate agents to flout? Countless ferry trips which I embarked upon with high hopes ended in bitter disappointment.

Coldside was described as being in the outer reaches of Cumbria. How accurate, being just four miles on the English side of the Scottish border. Intuitively, the moment I set eyes on it, I knew my search was over. My long trek had not ended in disillusion this time. I had found our home, our life and our future. We settled in and let the love that is Coldside embrace us.

2
The Orphan

Fate called again two years later, when the telephone rang just as Bernard was about to dash off and post a consignment of hedgehogs. These were not, I hasten to add, the live variety, but sweet-scented fabric ones with teasel or pine-cone heads that comprise part of the range of pot-pourri gifts produced by our little cottage industry.

'Hello, Marie, it's Amy, may I speak to Bernard?' I detected a note of anxiety in her voice.

'He's just leaving to catch the late evening post, Amy, but I'll ask him to pop in on the way back,' I replied.

As Bernard departed, I wondered why Amy sounded so discomposed, it was most unlike her. Eddie and Amy's dairy farm at the end of the lane leading from our house, forms one boundary of the Forestry Commission conifer plantation known as Coldside Wood, and is on the route to the express parcel depot eight miles away.

An hour later I heard Bernard's car returning. 'Daddy's back,' I called. Instantly our three girls rushed out, their tails swishing the air and mouths barking their greetings.

Bernard entered holding aloft a large straw-filled cardboard box, closely followed by three noses sniffing the air with the expectancy and excitement that heralded a new arrival.

Bernard carried the box into the utility room and placed it down on top of the freezer, away from nudging noses. Turning to the girls he said, 'We have a very special orphan who is too weak to be introduced today, so be good girls and go and lie down, nicely.'

Having ushered the reluctant girls into the lounge, and closed the door on them, I returned to the utility room. Bernard placed the box on the floor and gently parted the straw. My heart skipped a beat as I gazed upon our new arrival. How tiny she was. I instantly recalled the film *Bambi*, as I held a precious, fragile, roe deer fawn in the palm of my hand. She was perhaps two days old at the most, and so silent, so still.

'Where has she come from, Bernard?' My voice was heavy with misgivings.

'Amy found her in their barn, and assumes it was carried there by one of her dogs. She's unharmed, but can't stand, poor little soul.' She was obviously very feeble, probably hungry and maybe in shock. Bernard placed her tenderly back into the box.

Realising that we couldn't possibly attempt to rear her by guesswork, Bernard phoned our friendly vet for advice. Alas, his expertise did not extend to foundling baby deer.

It seemed the odds were stacked against us. Too many imponderables. We had only speculated as to her age. She could be younger. If so, she may never have ingested her mother's milk containing the antibodies so vital to a new-born to allay the onset of infection. Even if she had, enteritis was still a great risk. It can be caused by infection, and also by digestive disturbances, as is so often the case with a newly introduced diet. Although seemingly sound in limb, she could have internal damage. After all she had been in a dog's mouth.

Moreover, there was the shock factor. The trauma of

The Orphan

being removed from her natural environment, and subjected to much unavoidable handling, could bring about a respiratory problem and certain death. Putting it in a nutshell, he felt her chances were equal to that of us winning the lottery outright, and that perhaps a visit from him to administer euthanasia might be the kindest thing.

Bernard and I love *all* animals. Not the pat-on-the-head, walk-round-the-block kind, but a deep caring concern that doesn't transfer easily into words on paper but yet will be understood instantly by fellow animal lovers. Being presented with a living creature in need of help, our natural instinct to preserve that life, saw us clinging to the adage, 'Where there is life there is hope'.

Bernard told the vet he would phone back directly, after considering our options.

Returning her to the forest seemed a sensible idea at first, but as panic gave way to logic, certain questions needed to be considered and answered. From which part of the forest had she been taken? Forty acres aren't vast, but when one has no idea from where to start, it appeared to be an awfully large place. If we placed her in an area frequented by deer, perhaps she would be found by her mother, but if so would the smell of man jeopardise the little fawn? Had the mother had an accident?

Maybe Mrs Fox would consider the fawn to be easy pickings. After all, she too had a family to provide for. Was the fawn actually brought from the forest, or from the surrounding fells perhaps?

As we talked, Mother Nature was making the decision for us. I looked out of the window at dusk descending on the forest. Creatures and animals would be thinking about dinner, but our little fawn was not to be on the menu. Despite the vet's pessimism, Bernard told him of our intention to try and rear the orphan, and that we

BREEZE: WAIF OF THE WILD

would contact him should the ultimate option become necessary. Along with the vet's 'good luck wishes' was his valuable advice of giving nourishment little and often, very often, as with any new-born animal.

Darkness had now descended over the forest and this little waif of the wild in her straw-filled cardboard box was now totally dependent upon us for life, literally.

Unqualified as we were for such an undertaking, and with no inkling of what it might involve, Bernard and I determined to give this beautiful, delicate, fragile, willowy orphan our very best shot. So, on May 26th, we started our fawn foster parenting, and were not to know that we would be incarcerated in Coldside for an indeterminate length of time.

Our first priority was to give her sustenance. More for information than intent, we weighed her. She fitted easily on to the kitchen scales, and registered less than a bag of sugar. Bernard phoned Eddie at the dairy farm, and asked him with what he would rear an orphan calf.

Eddie said he kept a supply of colostrum in the freezer. This is the first milk drawn from a mother's teat — human or animal — and contains essential antibodies. He defrosted the colostrum as needed, and bottle-fed it warm to calves. He willingly gave us a supply.

We wondered whether her tiny deer tummy would be able to tolerate bovine colostrum. How much should we give her, how often, and with what?

My mind filled with foreboding. Sensing this, Bernard left the room and returned with an eye-dropper. Hardly a substitute for a mother's soft nipple, but ideal for the task in hand. He warmed up a small quantity of colostrum, opened her tiny mouth and trickled two eye-dropper's worth of nourishment down the back of her throat. Alien liquid apart, she had to accept us as 'mother'. If we didn't gain her absolute trust, we were lost . . . She offered no resistance to being picked up, but what about later?

The Orphan

'We'll see how she is in an hour, pet,' Bernard said gently. 'It's up to her now, we can do no more at the moment.'

As he placed her back into her box, she looked so forlorn, snuggled into a piece of simulated lamb's wool blanket.

The minutes ticked by agonisingly slowly towards the next hour. We peeped into her box, hardly daring to do so. She was still snuggled up in the blanket, fast asleep, and was breathing deeply and rhythmically. She was alive! I was consoled by this, and it gave me renewed confidence as I prepared her next dropper feed.

It was now ten-thirty, and the girls were becoming increasingly restless. The constant padding of feet pacing back and forth was now accompanied by soft whining. After a full day's outside activity, they usually welcomed their evening of rest. Not so tonight. For one thing they were not accustomed to being 'confined to barracks'. Besides, they missed our company.

Following normal routine, Bernard put on his wellies to take them for their evening 'empty bladders' walk. Their squeals of delight as he opened the lounge door and praised them 'Oh, what good girls you have been,' assured them that a closed door was not punishment for any conceivable misdemeanour on their part.

Having made their exit and return by the front door, directly off the lounge, so as not to pass the utility room which is en route to the back door, he then gave them their biscuit supper snack. He transferred the water bowl from the kitchen into the lounge, bade them 'Night, night', and closed the door firmly behind him.

Time again to re-fill the eye-dropper which Bernard gently administered for the third time.

Over a cup of coffee Bernard and I discussed the long night ahead of us. All thoughts of our planned craft-making were abandoned. The fawn was more important than money. Our first aim was to sustain her, but beyond this, we had no control. In order to give her every chance, we agreed that her feeds should be hourly day *and* night. This would necessitate us staying in the utility room where a watchful eye could be kept on her. At midnight her fourth feed was due, and we humans were also needing fuel to stave off our drowsiness. On tip-toe I entered the kitchen. 'Woof'! I almost shot out of my skin.

'It's only Mammy, girls,' I answered the guarding trio, 'go back to sleep.'

By the time I had made coffee, toasted the bread and spread the honey liberally, their sighs of frustration were replaced by gentle sounds of snoring. They had finally dropped off.

Throughout the dead of night, the eye-dropper was filled with continued regularity, and always administered with an apology. Upsetting for her as it obviously was dropping milk down her throat every hour, if we didn't, she would surely die from weakness. We hoped for the best, but we were also, as much as one can be, prepared for the worst. Logically, it was highly unlikely that she would survive. The vet's words flashed through my mind. 'Slim chance . . . Alien environment for a wild creature . . . Stress too much for her . . . Enteritis . . . 'Won't survive . . .'

The crucial question remained. Would her tummy tolerate the fluid? At this moment she still had a hold on life, and we clung to this fact.

The long troubled night at last gave way to dawn. Our bleary eyes looked into her box and saw her stirring from sleep.

Having agreed that twenty eye-dropper feeds constituted a suitable amount to judge its digestibility, Bernard gently lifted her up and we checked. *No* discharge! Her tummy had accepted the bovine colostrum.

My heart sang with an indescribable relief. At that moment her beautiful doe eyes opened and she looked up at me. I was transfixed. In that instant a love for her pierced straight to my heart. She had made her conquest.

My love story with this waif of the wild would now begin.

3
Pleased To Meet You

As we hugged one another in joy, our girls, aware of our presence, began their dawn chorus.

'Shall I take them out the front way?'

Bernard pondered my question before answering. 'We can't live our lives avoiding the back door, pet, nor keeping the girls confined to the lounge. If she is to be with us a few weeks, she must be raised as one of the family.'

'But the shock of the girls could frighten her to death, Bernard.'

'Well, there's no point in raising her for a few days, only to lose her later, when they inevitably meet.'

I closed the utility-room door and opened the lounge door to three vigorously thumping tails intent on delivering their usual 'Good morning'. What a wonderful indicator of happiness a dog's tail is – and how sad for dogs who have lost theirs to suit show-ring cosmetics.

As I put them out for morning 'pee-pees', I was aware of an extra display of excitement from them. They knew there was something quite special beyond that door, besides Bernard.

As they came back in from the paddock, they

hesitated outside the utility room, three pairs of pleading eyes begging to enter and be introduced to our new arrival.

They were shaking in anticipation whilst I trembled in trepidation as Bernard opened the door and gave the order to enter. 'One by one, thank you very much.'

A battle of wits and brawn now ensued for the privilege of going in first.

Jester won and bounded forward in her characteristically comic way. Not at all aristocratic as befits a great dane, but she was an absolute clown of a dog, hence her name. She was big, black and beautiful, with dark velvet brown eyes. The strong, silent type our Jester, and Dark Destroyer of all slippers. Also first-degree scatterbrain, given to staring dreamily into outer space – whilst standing!

'Nicely with the baby, Jester,' I demanded, as she nudged the box in her blundering way.

She looked up at me, her quizzical expression asking, 'Why is the baby voiceless?'

Being a dog of fixed ideas, her intrigue continued. As she had shared our lives for eleven years, I could read Jester like a book, and knew that her next move would be to raise an enormous paw and bring it crashing down on the box! She was instantly reprimanded with a firm, 'Leave'.

At this juncture Sheba seized her opportunity to examine the box at closer range. No papers came with Sheba describing her family tree, but her heritage was almost certainly German shepherd and great dane. Her beautiful head supported by her powerful and streamlined, short-coated golden body, long legs and extra-large paws made a unique combination. Ten years of age, her appetite for exercise was still insatiable. She would chase a ball and retrieve sticks endlessly, this *after* a five-mile walk, and then ask, 'Where are we going now?'

Pleased To Meet You

Sheba's intrinsic guarding instinct is at times very annoying. Her incessant barking – penetrating at best, ear-drum shattering at worst – is delivered at anyone passing by, be it human, bovine, equine, canine or feline. Actually I think she just loves the sound of her own voice!

Highly strung with a touch of feral in her, Sheba has an extraordinarily gentle and tolerant side, which was now in evidence as she gingerly approached, her eyes resolutely fixed upon the box.

Emma can sigh, and at this point Emma sighed her impatient sigh, which is several seconds longer than her tolerant one. Sheba, recognising this sound, reluctantly stepped aside, their pecking order long ago established, and made room for Bossy Boots.

Emma, about eighty-eight kilos of superlative mastiff superimposed on just a hint of great dane by courtesy of an errant ancestor, remained a puppy, a nine-year-old Peter Pan, and is affectionately referred to as Baba. Irresistibly beautiful, with the most expressive, loving brown eyes you could ever wish to see, Emma is the cause of more headaches in our house than any other twenty dogs you would care to name. Being acutely sensitive to raised voices had allowed her to 'get away with murder', and by the time she reached adulthood – *not* maturity – she was convinced that the earth revolved around her.

Her total lack of confidence makes her dependence and demands upon us total. She roars like a lion and charges like a bull at strangers, an undeniably effective deterrent, but should one be bold enough to place her directly under threat, she is reduced to mouse-like timidity.

One day I may write a cookbook entitled *Meals I Have Tried to Feed Emma*. Fastidious is not descriptive enough a word. No *table d'hôte* for Emma. If it's made for dogs and comes in a tin, Emma won't have it. Fresh

BREEZE: WAIF OF THE WILD

meat for dogs? Don't want it. Nice fresh meat intended for human consumption only, thank you very much.

Bernard and I are vegetarian. This is not because of Emma, I hasten to add, although a case for us turning to vegetarianism because of her could be made. This situation amuses our butcher. Two non-meat eaters purchasing prime cuts for a woofer! Emma does have a considerate side though, like permitting Bernard and me to live in the same house as her!

'Sniff, sniff, grunt, grunt.' Emma's frustration was showing. Emma has a preoccupation with things that smell. She recognised human smell, dog smell and assorted other smells, but this one was new to her. As she continued to sniff curiously there was movement from whatever was within the box. Her brow wrinkled in bewilderment as the lid of the box was slowly opened.

The moment she sighted our foundling, Emma grunted with delight. Emma was smitten. And with laughing eyes and wagging tails, the three girls welcomed this tiny, silent innocent into our family.

Expecting some reaction, we searched the orphan's eyes. Not a flicker of fear registered. She seemed to know she was in no danger from the girls.

Waves of melancholy washed over me as they fussed over their new friend, and I could not help looking towards the door as if expecting another arrival.

Joy, our beloved labrador, and first doggy rescue, would not be joining us. Affectionately known as Poddle because of her wiggly walk, our fourteen-year relationship had ended three weeks earlier, yet part of me had not as yet accepted this. Her very name encompasses all that she was to us. Her uninhibited zest for life, viceless nature and eyes shining with fun, made all of those who met Joy fall in love with her.

She had a passionate delight in water and all things very smelly. She would retrieve stones from the river for Bernard, shake herself, and then roll with great pleasure in a mound of the nearest noxious substance. Dear Joy. The essence of life. True, faithful, loving. You kept your tired old body going as long as you possibly could. We love and miss you.

How Joy would have loved to have met our foundling.

I felt the warmth of the climbing sun as the girls and I left Coldside for our usual morning walk across the fields. I stopped to say 'Good morning' to Joy as I passed by the paddock. In my secret moments I still hear her voice.

How golden her resting place looked today, with a mass of meadow flowers and the gorse now in full bloom yellow, sunshine yellow. Joy reminded me of sunshine, and was true to her astrological sign, Leo, sun-loving with a pure heart of gold. Is it coincidence or fate perhaps that Joy and I share the same birthday, August 4th?

The painted sun on her headstone seemed to generate warmth as it absorbed the rays of the real one. But this suddenly gave way to an abrupt freshness. Dandelions danced and buttercups bowed their heads in deference to this unexpected warm breeze, its whispering echoing in my ears.

I looked heavenwards, and knew that our waif of the wild was to be named Breeze.

4
Fur, Feathers and Foundling

Whatever the weather, a long morning walk had become part of daily living for the girls and I. Living in the rain-barrel of England meant that beautiful sunny days were savoured.

Today, the sun shone brightly. It was time to put out their sun-loungers, which were three single-bed mattresses purchased second-hand from a sale-room. When placed side by side facing the forest, the girls delighted in lying on them, absorbing as much sunshine and vitamin D as possible.

Usually they made a dash through the lobby to the garden as if the last one out wouldn't get a mattress, but this morning they made a detour via the utility room for a progress report on their new friend.

Bernard was occupied removing teasel heads from their stalks as we entered. 'Shh . . . baby's sleeping,' he whispered.

Satisfied that all was well with Breeze, they left to take up their sunbathing positions.

'I've phoned Amy and put her in the picture, love,' he said. 'I've also told the vet, who is delighted with Breeze's durability and suggests we continue with the hourly feeds of colostrum, at least for a few more days before introducing her to goat's milk.'

Our working day saw us stretched to full capacity from early May through August, and Breeze had arrived at one of our busiest periods of the year, with the Whitsuntide holiday about to begin. Having no idea at this stage how many hours we would need to devote to Breeze's welfare, we would just have to take it day by day, and hope to keep pace with our current orders.

Of course, as one or other of us needed to be with her at all times, we would need to revise our daily routine. My morning was spoken for with household chores, Bernard's in many outdoor tasks. Our afternoons and early evenings were occupied with the need to earn a living, the making of our crafts.

Our fawn-watch shifts were allocated, the mornings to Bernard whilst crafting. After lunch I would continue feeding and keeping an eye on her, crafting in the utility room, allowing Bernard to complete his outdoor tasks. Our craft evenings together would remain unaltered, except that we would do it in the utility room as opposed to the lounge. Our bedtimes seemed to have been cancelled for the time being.

I plugged in the coffee percolator, switched it to strong brew, and searched the emergency cupboard for a box of glucose powder. Another demanding day lay ahead!

Warm sunny days being so precious, they were not to be squandered on indoor chores, so I decided to bag up my loose pot-pourri outside in the garden. As was often the case when the sun shone, I deferred my housework until evenings. Baba grunted, then sighed as I asked her to move along and make way for me on her sun-lounger. As I worked, I delighted in the singing of the woodland birds and the industry of the bees. How vocal the hens sounded this morning.

'Hens,' I screamed, rushing madly towards the hen-house. I had forgotten to let them out!

'Sorry, chuckies,' I apologised, watching them waddle out just as quickly as their geriatric little legs would allow.

Only five remained now of the dozen or so refugees from a battery existence. The very occasional egg they proudly presented gave us more pleasure than a hundred from any other source.

The same couldn't be said of our pigeons, our prolific egg-laying, chick-rearing pigeons. Initially, one lost racing pigeon took up residence with us, eventually finding a mate. They had babies, who had babies, who in turn had babies. Unworldly in the ways of pigeons, we soon had over forty of them until someone knowledgeable suggested that we started piercing the eggs with a needle. We were assured it would not be distressing to the birds, and would most certainly put an end to the population explosion. Can you imagine having everything that wasn't indoors covered in pigeon droppings? Our guttering, vehicles, wheelbarrows, windows and washing all bore the trademarks of our feathered friends.

Eventually we managed to bring the situation under control. The pigeons delighted in playing a game with us called 'Guess where we've laid our eggs today?' With minimal nest-building materials – a piece of straw and a couple of twigs seemed to be the norm – they are capable of home-building in the most inaccessible of places. Thus much of our time is spent in nest-hunting. However, they live with us and have to be cared for so, having fed the multitudes, I returned to my craft-making.

Every hour on the hour, I slipped indoors for a progress report on Breeze. No signs of distress, and sleeping peacefully between feeds. Any misgivings I may have had as to delayed shock began to evaporate as the morning wore on.

Only noon, and I was flagging already. Time to search

the larder for some high-energy food.

'Carbohydrates, carbohydrates,' I said to myself, scanning the shelves. 'Ah, pasta, that seems about right.'

Lunchtime for Breeze and ourselves over, it was now Bernard's turn to take the girls for their constitutional before making a start on his tasks. First on the agenda was the chopping of sticks and sawing of logs for our wood-burning stove, neither of which appealed today. Experiencing that 'morning after the night before' feeling, he didn't much relish the idea of swinging an axe, but life must go on!

I felt a deep flush of contentment as I fed Breeze at due intervals, and sighed with relief at the unsoiled straw. As I sat vigil I wondered if deer were silent creatures. She had not as yet emitted even the merest squeak.

I suddenly felt myself drifting, needing more caffeine to keep me awake. Although sleepless nights were not unfamiliar to us, having nursed each of our girls through post-operative spaying, on those few occasions there was always the promise of an early night and a good sleep. Today, I hadn't this comforting prospect to look forward to. The mere thought brought about a yawn.

As I swallowed my coffee in haste, Bernard popped his head round the door.

'How is she doing?' he enquired for the umpteenth time.

The pre- 'Feed us, feed us' walks with the girls, being the last of the day, were usually taken jointly, with Bernard and I using this opportunity to say 'thank you' to whoever deemed us worthy of being given so much of the world's bounty. We decided temporarily to cease these reflective walks together until Breeze could be left alone, so with 'See you soon, love,' I left with the girls, leaving Bernard to complete an urgent order which, due to my lethargy, remained unfinished.

On occasions, hearing us return, Bernard would hide from us. Thus began a game the girls loved, one called, 'Where's Daddy gone? Go and find Daddy.' No room in the house was left unsniffed in their excited endeavours to locate Bernard, and the delight they expressed on finding him is a memory I will treasure forever.

'We're back,' I announced, entering the utility room. No answer, no Bernard. 'Where's Daddy?' My question spoken aloud had the girls dashing off on their favourite game.

My heartbeat accelerated as my eyes travelled towards Breeze's empty box. Waves of negative thoughts overcame me. She has had a relapse. She has died. Morbid images now crowded my mind of Bernard burying Breeze.

However, all was dispelled when Bernard, chaperoned by the girls, reappeared, cradling Breeze in his arms. Tears of relief filled my eyes as Bernard spoke.

'I've been for fresh bedding ... No, she hasn't diarrhoea,' he added quickly, noting my look of abject horror. 'I just thought she could do with a nappy change, considering all the liquid she's had, but her straw wasn't even damp!'

'I wonder if deer have to stand to induce urination, Bernard?'

'Well, there's only one way to find out, Marie.'

This was to be her first glimpse of the outside world since coming to us. Bernard gently placed her on to the grass. She lay quite motionless.

Her sandy infant coat, dappled with white spots, long velvet ears and doleful black eyes encapsulated her beauty. She lifted her ebony black nose, which twitched at the fresh air, and then slowly, falteringly, raised herself upon spindly giraffe-like legs. Another sniff in the air, and Breeze tottered drunkenly but determinedly towards the forest.

'Peep, peep,' she uttered, bird-like. 'Peep, peep.'

I was taken aback, as her voice, reminiscent of one of those squeaky toys I had bought for the girls, was not at all how I had expected a deer to sound.

We stalked her gingerly and, as she became aware of our presence, she froze and dropped to the ground, feigning death. Her instincts had taken command against us, the predators.

I gently turned her to face the house but she was insistent in her pursuance of the forest, her plaintive peeps filling the air.

The first clump of rushes she encountered met with her approval. She curled up small and welcomed the cloak the rushes would give until her mother returned to claim her. I tenderly gathered her to my breast, and carried Breeze back to her alien home.

Although Breeze had certainly declared that she had a voice, the purpose in taking her outdoors was to see whether she would urinate. She didn't! Why? What if she had a blockage in her urinary tract? I was now panicking slightly.

'Think positively,' I told myself over and over. Gradually I regained control of myself.

Her vocal signals had not gone unheard, and almost as her tiny cloven feet touched the ground she was surrounded by three canines. I resisted my impulse to intervene at this stage, knowing full well that in order to raise Breeze successfully, I must have implicit trust in our girls. Nevertheless, like a vigilant guardian I watched, tensed, expecting at any moment her body to be squeezed like a dog's squeaky toy.

Breeze stood her ground quite unperturbed. Any uncertainty I had was instantly dispelled as they ignored her front end and busied themselves in digesting the signals being transmitted from her anal area.

Time now for Emma to re-establish her dominance over Jester and Sheba. They, wise in the ways of Emma, backed off in time-honoured fashion.

Fur, Feathers and Foundling

Breeze hadn't moved during the canine investigation of her nether parts; nor had she 'peeped'. Now, however, perhaps by being just one to one with Emma, they began anew. Were these distress calls?

Emma immediately withdrew her nose from one end of Breeze and focused her attention on the other. This she accomplished in one short movement of her neck, and now Emma's huge head, hung low, confronted the source of the sounds.

Breeze remained motionless as Emma's tongue tip gently caressed her nose. Emma's maternal instincts burst into life, and a second, third and fourth kiss followed in rapid succession. Slurp, slurp, slurp. The power and passion of their delivery knocked the little fawn sideways.

'Nicely with the baby,' Bernard spoke sharply.

Adjusting her stance a little, Emma now licked the top of Breeze's head, along her neck and shoulders and down her back, all in one continuous movement. This obviously had a soothing effect on Breeze as her 'peeping' had ceased. Emma now stuck her nose beneath the vestige of the fawn's tail and began licking her genitalia. Breeze's fragile hind legs parted slightly, and she began to urinate.

I felt a sudden surge of elation. 'Bless your little heart, your very first pee,' I cried.

Jester and Sheba, hovering patiently in the background, were now allowed to share Breeze's company once more, Emma having fully established rights of ownership. But after two or three minutes of various parts of her anatomy being nose-nudged, Bernard felt that Breeze had undergone sufficient attention for the time being.

'That's enough, girls. Off you go inside now.' Reluctantly they obeyed. We followed them in.

Our lounge led directly to the front garden via a lobby. Individual Dunlopillo beds permanently adorned the

BREEZE: WAIF OF THE WILD

lounge floor: Jester's and Sheba's were beneath the window and Emma's was along the opposite wall, facing out. We humans were allowed to have any remaining space.

Breeze stood alone on the grass now, and Bernard and I watched her from the lounge window. After a tentative glance around, and with all the skill and precision of a novice ice-skater, she entered the lobby, then the lounge. She looked first in our direction, and then towards Emma. Uttering a half-hearted 'peep', she daintily sauntered to Emma's bed and curled up beside her.

Emma had allowed no-one, but no-one, to share her bed before, but now she gently licked Breeze's head as if she knew what tender little creatures deer are. A new softness appeared in her eyes as Breeze snuggled into her body and fell fast asleep.

In only twenty minutes Breeze had seen the outside world, found her voice, done a runner, had her first pee, and found two aunties and a surrogate mum. Sleep well, little one, and welcome to our family.

'Dinner, dinner,' called Bernard from the kitchen. Jester and Sheba bounded in to partake from their respective bowls, whilst Emma awaited table service.

We were reluctant to wake Breeze as she was sleeping so soundly, but it was almost seven o'clock, time for her dinner as well.

I gathered Breeze in my arms and gently fed her. How peaceful, how soft, how undemanding she was. I snuggled her down in her box to sleep once more.

After our meal of sugar and starches washed down with strong, black coffee, we busied ourselves in craft-making, trying to catch up on the day's lost production.

Now that the lounge door was no longer closed on the girls, they had unhesitatingly taken to their beds and were resting.

I looked across the utility room at the laundry basket,

brimming over with washing still to do, the pile of ironing lurking in the corner, the hoover waiting to be switched on and the kitchen scales unused since Breeze's weighing. How I wished I could join the girls.

The plaintive bleating of a lamb suddenly caught my attention. From the window I could see it running every which way, announcing 'I'm lost', enquiring 'Where's my mammy?' My eyes scanned the pasture in response to its plea.

The surrounding fields belong to Mitch, our friend. His farm is situated twelve miles over the border. Coldside once belonged to the Mitchinson family, grandfather having sold only the house twenty years prior to our predecessors.

Sheep predominantly graze the land about, with the summer addition of a few cows. We enjoyed the freedom of his fields and shepherding of the animals in between Mitch's daily attendances, informing him by telephone of anything untoward that might need his immediate attention.

'Bleat bleats' merged with 'Bah bahs.' The lamb was reunited with its mother, who now consoled her offspring by offering her teats as a pacifier.

'Bernard, how long is it before a lamb is weaned and becomes self-sufficient?' I asked.

'About four or five months, I think. Why?'

'Oh, I just wondered if a deer would follow the same pattern,' I answered.

'I shouldn't think so. Being wild, I'm sure nature would provide a much earlier weaning. I'm contemplating going into Carlisle library in the morning to borrow some reference books on deer,' Bernard added. 'We really ought to have a guideline to rear Breeze by.'

I agreed wholeheartedly, feeling somewhat relieved at this suggestion. After all, Breeze had arrived so abruptly, and however extensive our knowledge was on raising puppies to adulthood, a deer just isn't a dog.

This was a fact of which I was becoming fully aware.

Having tended to our little one, Bernard wearily plodded on with the craft-making whilst I dragged myself into the kitchen to do my chores. I switched on the oven, intent on baking after I'd done the washing. Hoovering I'd decided was definitely off the agenda for tonight, and the ironing I would do in the utility room later. By the time I'd finished the washing I felt 'all at sea' and thought better of the baking, switching off the oven, knowing full well I'd only end up doing a King Alfred. Not even the hens would say, 'thank you' for burnt cakes.

As Bernard prepared Breeze's nine o'clock feed, I topped up the caffeine supplier before accompanying him as I ironed. After scorching a second shirt, endeavouring to iron through closed eyelids, I gave in.

'I don't know how you feel, Bernard, but I can't see me lasting until dawn. I need a nap.'

'Like you, pet,' he yawned, 'I'm going to need an hour's kip.'

Considering it normal to spend a third of one's day sleeping, we certainly weren't getting our full share. To facilitate the next plan of action a camp-bed was brought in along with the alarm clock. This would enable one of us to grab an hour's shut-eye whilst the other remained on call. The alarm would remind us that Breeze's feed was due, should the watcher uncontrollably nod off.

After a day of excitement following last night's broken sleep, the girls were still out to the world when Bernard's pre-bedtime call of 'pee-pees' sounded. Having taken an age getting themselves together, they were in and out in two seconds flat, and asleep before their biscuits.

After our calorie-fortified supper and compulsory coffee, Breeze was given her midnight feed and the alarm was set for one sharp. The night shift was beginning again. As Bernard was a bit of a night owl, a hangover from his show-biz days, he was to plod on with the crafting, allowing me the first hour's rest.

As the wee small hours ticked by, each of us just managed to stay awake on our watch, kissing the other awake and switching off the alarm before it had time to wake the whole household.

Before collapsing on the bed for my final nap, I entered in my diary, 'Tomorrow will be Breeze's third day with us.' Then I said my prayers.

5
The Survivor

Tomorrow dawned another beautiful day. The sheer joy of Breeze being not only alive, but kicking, outweighed all lack of sleep. Bernard lifted her out of her box. The morning sunlight alighted on Breeze's ear tips, which twitched as I gently stroked her head and murmured, 'Mornin', Breeze.'

The partially open door was suddenly flung wide, and Emma charged in with Jester and Sheba in hot pursuit, almost knocking me off my feet. I was invisible this morning. My 'Good morning, girls' fell on deaf ears. I was obviously inaudible, too.

All eyes were fixed on Bernard cradling Breeze in his arms. 'Shall we go and do "pee pees" girls? Breeze can come too.'

Their expressions couldn't have conveyed more eagerness.

I followed the procession into the paddock, whereupon Bernard placed Breeze down. As soon as her feet touched the ground, Emma's nose was up her bum, and instinctively she licked her genital area. Breeze's suggestion of a squat, accompanied by urination, brought forth my appraisal, 'Good girl,

BREEZE: WAIF OF THE WILD

Breeze – oh, what a good girl.'

Emma proudly licked her baby's head, then nudged her gently. Breeze falteringly moved a few steps and halted. Emma nudged her again, and began walking alongside her in unison. It was apparent that Emma was encouraging Breeze to accompany her. Taking on the responsibility of mothering Breeze seemed to come so naturally to her.

'I'll go and put the toaster on and let the chuckies out whilst you supervise the training session,' I said, prising my eyes open.

Having breakfasted, Bernard fed Breeze before his twenty-mile journey to the library. As Breeze was content to sleep between hourly feeds, she wouldn't know if I slipped out for a short while. I decided I would take the girls for their morning walk as usual. And before departing, I peeped in on Breeze.

Unusually, she wasn't sleeping, and in fact was doing her utmost to jump out of her box. 'Stay!', the command I would give to our girls, wasn't having any effect whatsoever, so I made a hasty retreat, closing the door firmly behind me.

Twinges of uneasiness accompanied me during our walk, and I felt relieved as we crossed the stream and re-entered the paddock.

The girls sat in the out-house waiting to have their feet dried. Having been brought up to respect their home, they automatically do this even when they are clean and dry, like this morning. Normally I would go through the motions, keeping to routine, but I was not concerned about cleanliness this morning. My mind was elsewhere, on Breeze's welfare.

I left them all sitting, paws flailing at the air awaiting contact with a towel, whilst I hurriedly opened the utility-room door. My heart lurched, as my eyes travelled to the empty upturned box. Straw was strewn everywhere, and there was no sign of Breeze.

The Survivor

As I frantically searched behind the settee, the freezer, then the fridge, followed by the washing machine, and finally the tumble-dryer, I remonstrated with myself. The very first time I had been left in charge, I'd lost her. How was this possible? I had closed the door behind me when I left, and just opened it now. She must be somewhere in the room, I told myself, but where?

The girls, fed up with waiting for the drying session, now entered the room, unhesitatingly going to Breeze's box. Jester nosed the straw, snorting as she tossed it about, whilst Sheba's head disappeared inside the empty box. Emma was standing quite motionless, ears pricked, as if listening.

'Where's Breeze, Baba?' I enquired, beseechingly.

Her response was immediate. She hastened towards a large wicker basket piled high with logs, between the freezer and the settee, and grunted.

'She isn't there, Emma, I've searched behind it already,' I said.

But Emma would not be deterred. Her eyes were riveted *on* the basket. Beating her tail furiously, she proceeded to grunt, declaring defiantly, 'She is!', then buried her nose *into* the log pile.

A feeble 'peep' came from within the basket. Emma withdrew immediately, her big brown eyes full of anxiety, begging me to retrieve her baby from this forest of pine logs.

As my eyes scanned the basket for a sign of her, I was suddenly knocked sideways.

'Jester, *NO!*' I yelled. It was too late. As I spoke her nose had bulldozed the log pile.

'Get out!' I screamed angrily, as logs toppled in all directions. 'Out, out!' My voice, rising in hysteria now, had them all beat a hasty retreat, for they knew better than to ruffle my feathers.

'Tring, tring' – it was the telephone. Well, it would

just have to keep ringing. I certainly wasn't about to answer it, I had a crisis on my hands.

I turned my attention to the log basket, surveying Jester's destruction. 'Where do I start?' I asked myself. It was no good deliberating, I would just have to remove every log until I reached Breeze. Slowly, and ever so carefully, I began lifting each one, listening all the while for the slightest whimper, but there was none.

The only sound I could hear was my heart pounding. Suppose Breeze had suffocated under the weight of the logs? I felt a tightness gathering in my throat at this terrifying thought, and my eyes brimmed involuntarily.

My trembling hands abandoned caution, as I picked up the logs with increasing urgency. Through blurred eyes I saw her, lying motionless, eyes closed. I took a deep breath and snatched her quickly as the surrounding logs all tumbled in. How cold she felt. I hugged her tightly to my chest, letting my touch warm her.

'Oh, Breeze, please be alive,' I implored, my voice beginning to break now, as the band of tension around my head tightened. I tenderly kissed her forehead and made a silent plea. Her eyelids flickered as she grew perceptibly warmer, and I could feel her chest rising and falling with tiny, even breaths. My prayer had been answered.

I placed her down on to the soft cushions of the settee, now swathed in sunlight, letting the warmth of its rays seep into her tiny body. As I sat alongside her, watching her sleep, thoughts of Joy returned.

This settee had been purchased for Joy. In her dotage she sought privacy, and retired to the utility room. Not wishing her to be in draughts, we scoured the second-hand shops in search of a comfy armchair, and returned with this two-seater settee. How her beautiful eyes lit up when we announced, 'This is Poddle's bed'. Understanding every word we spoke, she immediately

The Survivor

jumped up on it, and laughed her approval. All her resting periods from that day were spent on this settee. I could see her now, head resting on the arm, snoring away.

Since Joy's departure, none of our girls had been interested in occupying the settee.

How irresponsible of me to leave this little innocent, completely unaware of danger, for one minute let alone thirty. She could have died an agonisingly slow death if I had been out any longer. I felt a cold lead weight in the pit of my stomach. Having upbraided myself, I vowed *never* to leave her unattended again.

Suddenly cold and drained of vitality, I tip-toed out to the kitchen to make a strong, hot cuppa. Having stirred in a spoonful of honey, I steadied the cup in the saucer and returned post-haste to Breeze. I did a double-take. Emma was alongside Breeze on the settee, showering her with kisses.

It was touching to see her. I thought that I had never witnessed anything more beautiful. Emma's tail tip lifted gently as she saw me, and her eyes were pleading.

'It's all right, Baba, you can stay with your baby,' I reassured her. 'You love her too.'

My sip of hot, sweet tea was interrupted by the sound of a car pulling into the yard. Sheba and Jester's excited dash to the door told me it was Bernard. Strangely enough, all our girls could differentiate between car engines. Jester's paws had begun marching up and down the door before I had time to put my cup down.

'All right, all right, let me get the door open first,' I implored.

Like greyhounds from a trap, they raced to greet him, almost taking my legs from under me.

'Hello, my Jester, hello, my Sheba, and where's my Emma then?'

'She is comforting Breeze. I've had quite an ordeal since you left.' I babbled on for some minutes relaying the morning's events, in a voice I hardly recognised as my own, it was so shaky.

Bernard tenderly put his arms around me and drew me to his chest. 'There, there, it's over now, let's go and tend to the "little one".'

Emma's tail thumped her delight at seeing Bernard, and she kissed him repeatedly as he knelt down in front of her, stroking her with his left hand and Breeze with his right. Jester and Sheba pushed their heads through the crook of his arm and sniffed Breeze. As Jester's nose disappeared into Breeze's ear, a large growl gathered in Emma's throat. Her devotion was manifesting itself in extreme possessiveness as she lay guarding her baby. They got the message and retreated immediately.

Breeze's sleepy eyes opened, and she began licking her lips.

'Come on, girls, outside in the sunshine, it's time for baby's feed.'

Jester and Sheba followed Bernard out, but Emma had decided that this order was not directed at her, as she didn't move an inch. Bernard returned with an egg cup of colostrum and the eye-dropper. After quietly observing Breeze's curious way of partaking of 'dinner', Emma snuggled up to her once more.

'I wouldn't say no to a cuppa, Marie, I'm parched.'

'Me too. Would you like a sandwich?'

'I'm not really hungry, pet, a piece of your oatcake will do.'

I admitted to not having any appetite myself, as we adjourned to the kitchen.

Gulping down his strong black coffee laced with sugar, Bernard extracted a lone book from his bag. 'This is it, pet! The only one the library have on the roe deer and its life in the wild. And I'm afraid there isn't any information on how to raise a foundling deer to maturity by hand.'

My heart sank. I had hoped that someone,

somewhere, must have successfully raised an orphan and written about it. Not so. Now what?

Guessing my thoughts, Bernard took hold of my hand. 'We have done all right so far, Marie. Don't worry, Breeze is a survivor.'

I forced a smile to dispel my worried frown.

'That's better! Now get your pen and paper, and we will extract all the pertinent material to work out a formula for rearing her as close to nature as we can.'

'Let's sit outside in the sun. I feel so cold and shivery,' I suggested.

'I'm not feeling too warm myself,' replied Bernard, 'probably due to lack of sleep.'

I agreed. After all, the human body will not function for long without proper fuel. It needs many things, like sleep, and we were most certainly not having our full quota.

I donned my straw hat and we joined the sunbathing duo in the garden. The sun was shining out of a clear blue sky, and the scent of the lilac now in full bloom filled the air. It certainly was a beautiful day.

Pen and pad at the ready, Bernard started to read. As the facts unfolded, the realisation of the enormity of our task dawned on me.

6
Facts, Fears and Fatigue

The roe is not a social animal, unlike herd species such as red and fallow deer, but is solitary. It is the smallest of our feral deer, adults standing slightly over sixty centimetres to the shoulder. They are territorial, the buck holding his established territory against newcomers, only breeding does being allowed to share this area with him.

Usually, roe have twin kids. Immediately after giving birth the doe swallows the placenta and membranes, licks the kid and leaves it, moving three to four metres away. When she drops her second youngster, she repeats the cleaning process, before pressing each one down with her nose, being their 'Stay!' command.

They remain apart and completely silent, concealed in undergrowth, whilst mum goes off to feed on her own. She remains within earshot and returns to suckle them independently every four to five hours. Leaving them separated until physically developed enough to follow her is designed to protect them from predators. Although we no longer have wolves in Britain new-born fawns are still in peril from foxes and roaming dogs.

Somewhere, Breeze's twin had lain undetected.

At about three to four weeks of age they are introduced to each other and then lie together as a family unit. From this time onward they forage with their mother, learning which grasses, herbs, fungi, shrubs and fruit are edible.

'Learning?' I asked, incredulously. 'I had assumed that she would naturally eat all herbage, like a lamb, didn't you, Bernard?'

'Well, yes, but according to this book, deer cannot tolerate certain plants and bushes, whilst privet and rhododendron are fatal. Haven't you just planted a rhododendron?' he asked innocently.

'Yes, but how was I to know it was poisonous to deer?' I responded defensively. I cursed myself for not knowing. How on earth were we going to educate Breeze correctly?

I had never observed a deer eating at close range. Hardly surprising as roe lead very secretive lives, preferring to forage under cover of darkness. I wasn't blessed with exceptional night vision, and Bernard's visibility in the dark is nil.

The thought of crawling around the forest floor in the pitch dark was daunting. I had visions of me emerging at dawn from the tangled undergrowth full of nettle stings, pine needles up the bum, lumps and bumps from midge bites, and in a stupor through lack of sleep. A state I wasn't far from now!

My pen trembled as I tried to focus on my pad, whilst Bernard continued informing me of a deer's diet.

Roses, bilberries, ivy, bramble and male ferns are favourite food plants. My eyes alighted on our rose bushes, then the ferns (whether or not they were male, I wouldn't know). We certainly hadn't any brambles, ivy or bilberries, but I could plant some. Perhaps we could encourage her to use the garden as her larder. I would have to find a kind home for the dreaded rhododendron, though.

A nightly walk in the garden I wouldn't object to at

all. I could even leave the door open and sit inside the lobby, sewing. Perhaps if Breeze could see me, she would feel secure enough to munch on her own.

Apparently all ruminants need to have a steady supply of food passing into their rumen, otherwise the digestive system becomes disturbed. Well, the garden was in full bloom, plenty of food here to fill her tummy. After all, it will only be for a few weeks, and an ever-open door during the summer wouldn't be a problem at all, I told myself. Bernard could actually get a full night's sleep, and maybe I could doze off for a few hours during the day. My midnight mingling with the woodland's wildlife worry was solved. Bernard carried on reading.

I frowned with the beginning of a headache and lowered the brim of my hat, as the sun shone with increasing strength. The next words I wrote are embedded in my memory forever, for they were to alter the whole course of our life.

The bond between the doe and her fawns persists throughout the summer, winter, and the following spring. In fact they stay with their mother until about a week before the arrival of her new offspring. This is when the doe *drives* off her yearling fawns, and so the family group is split up. This may take many days, because the youngsters are reluctant to forego her company and protection. Thereafter, the yearlings must go in search of their own territory.

Breeze was to be with us for *twelve months!* I had thought that we would only be fostering her until she was weaned, after which she would merrily trot off into the woods to join her own kind. How naïve I was! We were totally unprepared for such a long relationship, but to contemplate casting her out when she attained her first birthday filled me with despair.

I suddenly felt that the day was a bit less glorious. I had butterflies in my stomach, and a ringing in my ears.

'There's the phone, Marie. Hang on while I answer it.'

'Bernard, will you bring two fizzies back with you, please?' I asked, clutching my head.

'Of course, love, I need a couple myself to be honest,' he replied, wearily.

Fizzies, as we refer to them, are those effervescent type of headache tablets that act quickly. I really was in need of an instant panacea, like sleep. My ear-buzzing ceased as Bernard picked up the telephone.

As I sat registering the implications of all we had read, Bernard returned with a glass of bubbles and an urgent order for 'smellies' from a hotel in Gretna Green. Having sold everything over the public holiday, they were besieged with visitors and in need of stock. Since Breeze's arrival, we had managed to make a grand total of sixteen hedgehogs, had done very little housework, and the outside tasks were building up.

'How are we going to afford the hours necessary to devote to Breeze, as well as trying to earn a living?' I asked pleadingly. I could see the ill-concealed anxiety in Bernard's big blue eyes, dull with lack of sleep, as he endeavoured to reassure me.

'It isn't going to be easy, pet, but we'll make the time somehow. Now off you go to Gretna, the drive will do you good. We'll discuss Breeze's upbringing this evening. Don't worry, we will work something out.'

In spite of my resolve to relax and enjoy the eighteen-mile journey to Gretna, my thoughts kept dwelling on Breeze. The questions swirled and expanded. How were we going to find the time needed to see to Breeze's needs twenty-four hours a day?

Breeze's need to follow her natural diet was certainly going to cause us lots of headaches. Just thinking of them brought my current thumper to the fore. I bent my head towards the open window and welcomed the sign in the distance, 'You are now in Scotland'.

Facts, Fears and Fatigue

Armed with boxes, I entered the hotel. Freda's smile lit up her face as she greeted me, voicing her surprise.

'Hello, Marie, how did you know we needed replenishing? You must be psychic.'

'Didn't you speak to Bernard about an hour ago?' I enquired.

'No, I rang at about noon, but there was no reply,' she answered.

It suddenly dawned on me that I was at the wrong hotel. Bernard had stated it was the Hunters' Lodge that rang, and here I was at the Royal Stewart.

'I'm ever so sorry, Freda, my head's in a bit of a spin today. I've actually come to re-stock Jan and Graham's hotel, but you can have half of what I've brought with me to keep you going,' I said apologetically.

'Overdoing it as usual, I expect. Come and have a cup of coffee.' There was always a percolator of fresh ground coffee on the go, and today out came a tin of biscuits, with a concerned comment on how I needed fattening up. Admittedly my normal weight had dropped a bit in the last few days. Little wonder!

'I'll have my coffee as I put the new stock on display – I'm pushed for time today,' I said.

'Suit yourself. I'll leave you to it. Help yourself to biscuits,' replied Freda.

I hurriedly refilled the depleted stand and returned my empty cup to the kitchen. 'Thanks for the lovely coffee, Freda. I'll call on you next week, unless you need anything before then, in which case ring me.'

'Okey-dokey. 'Bye, Marie, and slow down,' she urged, wagging her finger at me.

'I'll try, Freda,' I replied as I did the quick-step to my car.

I battled to find a space to park at Hunters' Lodge. The limousine adorned with white ribbon and the trail of confetti indicated Jan and Graham had a wedding party in. Not wishing to intrude, I exercised my diplomacy by ringing the bell at the tradesman's entrance.

'Hello, Marie, what are you doing coming to the back door? We were expecting you earlier. Jan's busy in the kitchen and I'm pouring drinks, but do come in,' said Graham cheerfully.

'I won't interrupt your timetable, thanks. Could I just leave the boxes? I've enclosed a list for your records. Would Jan display them for me when she has time?' I asked graciously.

'No trouble, Marie, we'll see to it later on. Take care now, see you soon,' Graham chirped, his broad smile sending me on my way.

The car clock stated five in the afternoon as I left Gretna Green, homeward bound. No doubt the girls would be hustling Bernard for their evening walk, having missed out on their usual after-lunch one. It would be getting on for six, though, before I arrived home, and knowing Bernard would not leave Breeze unattended, I drove in feverish haste.

Having been well exercised and eaten heartily, the girls were happy to rest. Not so the hens, who were enjoying the evening sunshine until my sweeping brush behind them and Bernard scattering digestive biscuits, their favourite treat, in front of them effected their eviction from the garden into the hen-house. Loud protests of 'It's not bedtime yet!' in cluckers language ensued as Bernard slammed the door shut. It's undeniable that all domesticated animals have a built-in time-clock which determines their daily routine. It is particularly evident when British Summer Time begins and ends. We humans can adjust and work to the altered hour, but it confuses animals.

Although we disliked having to lock the hens away prematurely, it was vitally important that this particular evening was made available, without disruption, to do some deep thinking and get down to devising a plan whereby we could raise Breeze to maturity.

Breeze, contented as ever, accepted her next feed with a lip-smack. Our taste buds weren't interested in food proper, but we cajoled our tummies into accepting a bowl of muesli drowned in yoghurt. 'A healthy body begets a healthy mind.'

We spread our data on the kitchen table. I was barely able to concentrate. The words were there on paper in front of me, but I couldn't see them, they were just a blur. There is no denying that constant worry and lack of sleep leave their mark upon the body and restrict its activities. I tried desperately to send a message to my fingers while, systematically, we dealt with the facts.

This task could not be hurried. We talked long into the night, discussing how best to rear Breeze. We had to introduce her to milk, then she had to be weaned, so vegetation came next. The garden would suffice for the time being, but it couldn't support her for a whole year, as everything dies back in winter. The ideal solution would be to site a caravan in the forest and for Bernard and me to take alternate shifts living there. At least we could work whilst remaining in close proximity, allowing Breeze to forage in her natural environment, but I couldn't see the Forestry Commission agreeing to this request. Forestry plantations are not grown for wild herbivores to enjoy, but in order to produce valuable timber. A deer's crime of eating to stay alive, of marking out its territory and cleaning its antlers against young trees interferes with man's interests. This is the reason of course, for the long-standing persecution of deer.

Her future suddenly looked bleak.

It wasn't only a question of food source, more importantly it was where she *belonged*. She needed to live as natural a life as possible in order to develop, and to equip her for her eventual life in the wild. But how, if she was barred from her natural home?

Bernard took my hands in his and, pressing them to his cheek, smiled supportively, as he gently spoke. 'Look, love, we agree it is cruel to let a wild animal live under the illusion that it is anything other than free, and no way will we attempt to rear her as a domestic pet. When the time comes for her to eat of the forest, we'll take her, and accompany her, and not say a word to anyone.'

Although this was now sorted out, we kept coming back to the same problem. How would we find the hours, day in, weeks and months out, to devote to her welfare? There were the nights to consider as well. Excursions to the forest would be necessary, allowing Breeze to follow her predominantly nocturnal instincts. Difficult as it would be, we were going to somehow have to find the time. Breeze's very existence depended on our doing so.

I bit hard on my lip to freeze my next words, but my questioning eyes were a give-away, as an expression of knowing spread over Bernard's face, now white with fatigue. We both knew we were avoiding the ultimate question. Would Breeze's very dependence on us prejudice her chances in the wild?

You have to be able to accept the answers to the questions you ask, so we both remained silent.

With Breeze's future upbringing mapped out, there was no room for complacency, however. Albeit she had survived three days, this did not guarantee a fourth – nor the next hour, for that matter.

As another night of vigilance gave way to dawn, our lives were well and truly centred around Breeze.

7
Breeze's Graduation

My tiredness had insisted I nod off in the middle of my fawn-watch. I awoke heavy-eyed to the sun's rays offering the gift of another lovely day. Intuitively my bleary eyes focused on Breeze's box. Empty! Panic instantly jolted me awake, dispelled my daze of confusion and engaged my brain.

Then, remembering Breeze had moved to new sleeping quarters, I turned to the settee and stared in astonishment at Emma's face, signalling maternal bliss as Breeze lay contentedly cocooned in her commodious body. I gently roused Bernard and silently pointed across the room. As we gazed, hypnotised by this scene of affectionate intimacy, the spell was broken by the intrusion of Jester and Sheba, whose eagerness to fulfil their morning ablutions hurried me outside.

I felt like a somnambulist floating across the yard, and I had reached the paddock before realising Emma wasn't with us. I turned back and called out 'Are you coming, Baba?'

I doubled up in hysterical laughter as Emma appeared, shunting Breeze, who resembled a ballerina on points about to lose her poise, across the yard. Having deposited her in the paddock and supervised her ablutions, Emma went about her own toilette.

Emma shepherded Breeze back to her room with an air of possessiveness that she couldn't quite conceal, before accompanying Sheba, Jester and me on our morning walk down to the meadow and along the river bank.

As we returned, Bernard was preparing Breeze's breakfast. This was to be her last feed of colostrum, for we had decided her graduation time had arrived. Surmising she was between twenty-four and forty-eight hours old when orphaned, we calculated Breeze had reached the tender age of six days old.

We had agreed to take one step at a time, dealing with each situation as it arose, and our next task was to introduce Breeze to whole milk. In her wild state her mother's milk would no longer carry antibodies, therefore colostrum would not serve any further useful purpose.

Whilst appreciating that a fawn wasn't a lamb, we decided to hand-rear Breeze as one would a lamb because it was a ruminant of a similar size. And no-one was more knowledgeable in this area than Mitch's mother, Jean. Sadly, many lambs are orphaned at birth, or simply rejected by their mothers, but Jean's expertise with a bottle and teat, tempered with patience, always produced outstanding results. Bernard left to consult with Jean, whilst I telephoned Mary to arrange a supply of goat's milk. It was the vet who had intimated this would be most akin to a deer's.

Bernard returned with a supply of lambs' teats and a bottle. 'Jean suggests we give her two fluid ounces warmed slightly, every three hours,' he said, 'and that we should increase it gradually over the next few days. Apparently, not all lambs readily suck, but you must never force the teat down the throat in case some of the milk enters the lungs, which could prove fatal. Sadly, some lambs totally refuse to be bottle-fed.'

A shockwave ran through my body as I cut off Bernard's words, 'Oh no, I couldn't bear to lose her now.' My throat was constricting by the second.

'Calm down,' Bernard implored. 'We are not going to lose her.'

'But she will starve to death if she refuses to be bottle-fed!' My voice trailed off as I gave way to tears.

Bernard's arms reached out to comfort me. 'There, there, now, don't upset yourself, she isn't going to die of starvation. There is an alternative way to feed her. You interrupted me before I could explain,' Bernard reassured me. 'Now listen, and don't panic. There is something called a feeding tube, which is inserted down the throat, directly into the stomach, and the milk is poured through the tube. Jean doesn't have a spare one, but the vet carries a supply. He could come out and instruct us on how to use it correctly, as it's an exacting procedure, but I'm sure that this won't be necessary. Breeze will readily take to her bottle, make no mistake about it, she wants to survive.'

Bernard's soothing words conveyed his unwavering trust in Breeze's strength of will.

'Come now, let me dry those tears,' Bernard said softly, patting my cheeks with a tissue. 'I'll make you a nice cup of hot tea, with honey in, then off you go and collect Breeze's milk. Stop worrying, we won't need to force-feed her.'

Bernard's words of assurance helped my mind to settle and find a balance once more. Whenever I felt less than level-headed, Bernard was my protector, the barrier between sanity and insanity.

Mary was just finishing the morning milking when I arrived. Her beaming face reflected her delight at seeing me. 'How thrilling to have the opportunity of rearing a fawn. You must be so very excited, Marie,' she brightly exclaimed.

I nodded agreeably, not wishing to go into the great debate of how 'exhausting' would have been a better choice of word. Was it only three nights we had gone without proper sleep, I thought to myself as Mary air-sealed the warm milk into a pint-sized bag.

'I've specially drawn it for Breeze from my prize-winning goat. Now I'm not sure how long a pint will last her, but just ring me when you require more.'

I thanked Mary for her good-naturedness, and her 'good-luck' wishes hastened me home.

The empty bottle in a pan of warm water was awaiting its contents. I watched Bernard lovingly preparing Breeze's very first bottle, continuously testing the temperature of the milk by allowing drops to fall from the teat on to the back of his hand.

'It's perfect now, pet, let's go and feed our "little one",' he said eagerly.

Lifting Breeze from the settee, I placed her tiny, delicate feet firmly on the floor. Bernard knelt in front of her and, tilting her head upwards in line with the bottle, gently inserted the teat between her lips and invited her to suck. Breeze drew her head back in silent rejection, declining the invitation.

Again he persuaded the teat into her tiny, tightly closed mouth, and gently stroked her throat. Breeze's reply was again a firm, 'No, thank you.'

My anxiety was growing by the minute as I watched Bernard repeatedly coaxing the teat into her unwilling mouth, and Breeze, in quiet protest, as repeatedly withdrawing. I sighed in helpless despair.

'Relax, Marie,' Bernard implored.

'Relax, how can I possibly relax?'

Bernard – so rational, so calm, so aware that Breeze's very existence depended on him – continued to encourage her to suck. By now every nerve in my body was jangling as Bernard, intolerant of defeat, kept persevering only to be spurned again and again. Lines

Breeze's Graduation

of tension were now appearing around his eyes and across his head. His face looked so drained.

'Come on, baby Breeze, please suck for Daddy,' he begged softly.

As if understanding, she gave a feeble pull on the teat. The room resounded with the sound of a swallow, then another, and another. The lines on Bernard's face evaporated along with the milk. The empty bottle said it all. We were over the first hurdle but not out of the woods yet.

In the light of the previous introduction to colostrum, the next few hours were to be crucial. If her tiny tummy could not tolerate the change in diet, all our efforts would have been in vain.

I banished the thought from my mind, occupying myself with gardening. An age seemed to tick by until Bernard called out, 'Where are you, love, it's time for Breeze's next bottle.'

I clasped my hands in supplication and held every breath in my body, as anxiety threatened to give way to panic. As I said, 'I'm coming', Bernard took my trembling hand in his and we entered the utility room.

Breeze arose instantly, daintily stepped down from the settee and trotted towards Bernard's outstretched hand. Her tiny tongue peeped out briefly, only to emerge again as if licking her lips in anticipation of her lunch.

I felt the blood leave my cheeks and my knees wobble as I forced myself to look at where Breeze had just lain. My anxiety was instantly dispelled. 'It's dry, Bernard!' I cried jubilantly.

A broad smile lit up Bernard's face. His delight was twofold, as Breeze, needing no inducement this time, guzzled her milk with gusto, to the very last drop.

I flung my arms around Bernard and hugged him. 'Thank you, pet, oh thank you,' I cried, as my voice cracked with emotion. 'It was the special ingredient you mixed into her milk that was the governing factor.'

BREEZE: WAIF OF THE WILD

Bernard smiled knowingly. 'That's why I hadn't any misgivings,' he admitted modestly. 'Haven't I always believed that all animals respond to love? Love is such a positive force in life, but it needs encouragement if it is to flourish.'

Breeze thrived on her diet of love. A love, however, that was to leave me defenceless.

8
Laying the Foundations

That evening felt like heaven. Three full hours of unbroken sleep spurred me on to an equal period of continuous craft-making whilst Bernard, having downed tools, took his opportunity to recharge his body's batteries.

The following day, with fewer breaks from our work, enabled us to improve slightly on our back-orders. However, as the evening approached, it became blatantly obvious that our awake-shift crafting would have to cease. We both needed less mental activity, less coffee, less junk food, and much more than three hours' sleep. We agreed that whilst we needed to be in the utility room to see to Breeze's three-hourly feeds, we would both bed down together after her midnight bottle until the alarm woke us at three to feed her again. After which, we would both return to sleep again until six.

Although both of us are of slim build, the camp-bed was not going to allow us on it at the same time, so we swapped it for the spare room's double divan.

The first three hours leading up to the alarm going off were most gratifying. Bernard fed Breeze, and I turned over to enjoy another three hours. As the first approach of sleep touched my mind lightly, I began to dream, but the moment I dropped off it turned into a nightmare. I woke up with a start, and rushed over to Breeze, dreading that the horror of my dream had come true. The sheer relief of finding her still alive, blissfully sleeping, not only reduced me to tears, it opened the flood-gates to my mind. Although I had never ceased for one moment worrying about her survival, up until now I had kept my heart distanced, fearing a trough of despondency if she died. Now that she was 'one of the family', my emotions unleashed themselves. From then on, I slept very little, in fact not at all.

Three hours of worry as to her future well-being served only to give me a thumping headache brought on by nervous exhaustion. Needless to say, all this apprehension was without foundation. She couldn't have awakened any healthier, wetting her lips repeatedly as Bernard filled her bottle.

Of course there had to be a fly in the ointment. It was Breeze's refusal of yesterday's milk. It had to be freshly drawn. Even Mary's suggestion of freezing it whilst fresh then thawing it before use was not to her palate. And so began the twelve-mile trip to Mary's each day at dawn.

Throughout that day I busied myself in work, shutting all thoughts of her out of my mind. Because sleep normally follows fatigue, I had thought that as midnight approached, I would collapse into bed and sleep right through until dawn. Alas, the effect of my mental disquietude caused the very opposite. Yet again, I couldn't sleep at all. After this second night of tossing, turning, and unfounded worrying, my whole body felt wound up tighter than a reel of cotton.

In the morning, first and foremost was the collection of Breeze's milk. Secondly, at Bernard's insistence, I had a session of meditative yoga.

'It doesn't matter if you shut yourself away all day, Marie, you must rest your mind.'

Having by now almost reached a pitch of nervous prostration, I retreated to the sanctuary of our bedroom. It was three in the afternoon when I rejoined my family, relaxed and fully rested, having drifted into a serene, deep sleep.

'How is baby Breeze, Bernard?' I asked, gently stroking her soft head.

'Absolutely wonderful, pet. She's done precious little of anything but eat, sleep and pee,' he replied.

Over the next couple of days nothing changed as far as Breeze was concerned. The more she drank, the stronger she became, and the wetter the bed.

As Breeze's liquid intake increased to a quarter pint every three hours, so did my washing. I soon learned that a deer does *not* need to stand to urinate, and are, quite naturally, 'bed wetters'.

As the settee was now firmly Breeze's patch we had to do something to protect the base. A ground sheet proved to be the solution, over which we laid a layer of nappies given to us as 'dog driers', courtesy of my sister whose family had long outgrown them.

A line full of nappies blowing in the wind as if heralding a new arrival caused a few raised eyebrows in the locality!

In addition to a mountain of washing daily, there were bottles to sterilise and feeds to prepare. With our time swallowed up and craft orders to complete, little wonder therefore that the house was appallingly neglected. To be honest, it was a shambles with any semblance of normality long gone. Our days now revolved around Breeze.

Was it really only eight days since this tiny creature had arrived so unexpectedly, I asked myself as I reached for the honey, my energy booster, and spooned a dollop into my mouth.

Washing the spoon, I caught sight of myself in the mirror that hung on the wall above the sink. Did I really look like that! The strain that this particular week had imposed on me was most definitely showing. I actually looked as bad as I felt!

A penetrating bark almost shattered my ear-drums. Sheba's alarm call told me someone was approaching. Her barking increased in volume and intensity as the post van drew to a halt. It had to be Ann as it was only eleven. Our postal delivery by Bill, with whom she alternates, was always around noon. I snatched the only item of mail I had to go off today. My normal eight-page weekly letter to my mother was merely a postcard carrying a brief message. 'Sorry I haven't had time to write due to a new addition to the family. A *DEER*, not another dog. Will write more fully later.' As Mum lives in Darlington, and is not as yet on the phone, she looks forward to my letters and to relaying our news to my three sisters and brother, which helps bridge the ninety-six mile gap between us. I felt quite guilty at having written so few words.

Dashing out to catch Ann before she departed I felt a sudden surge of heat as if I had just opened the oven door. I hadn't realised it was so hot outside. The thickness of the walls in older houses stops penetration of the elements, wonderfully cool in summer and warmth-retaining in winter.

'Morning, Ann! Goodness, it's hot, it's going to be a scorcher today.'

'It's roasting inside this van, I feel like my engine, almost boiling over!' replied Ann, her flushed face bearing witness to her comments. 'Are my eyes deceiving me, or have I just seen a tiny deer sitting with your dogs in the garden?' Ann enquired, with a flicker of a smile.

'It must be Breeze,' I answered, 'but what is she doing outside? I left her curled up asleep on the settee.'

'Well, if Breeze is a deer, she is sunbathing, so that answers your question and mine,' Ann replied, her smile now a mischievous grin. 'And I'm relieved I won't have to book an appointment with the optician!'

Suddenly the humour of the situation struck me as I laughed aloud. How stupid I must have sounded.

After listening to my story of Breeze, Ann continued on her journey down the lane to her next point of call, our neighbours Ralph and Jean, and their sons Christopher and Ken. They occupy The Craigs along with Toni and Tani, their lovely old labradors, Mandy the poodle and about eighteen cats. Beyond The Craigs is the final residence, Cleughside, the working farm tended by George, his wife Dorothy, and his sister Essie, whose dogs, cats, peacocks, pigeons and hens seem to outnumber their sheep.

People in these parts are not known by their surnames but by the name of the house they occupy. I discovered this not long after moving here. In conversation with someone nearby, I referred to neighbours who had lived here all their lives, and found it strange that the person didn't have a clue who I was talking about. It wasn't until I was asked where they lived, and having stated the property's name, that the individuals concerned were instantly recognised.

Although Sheba's barking had ceased when Ann departed, I knew it was only temporary. It would certainly resume when Ann did an about-turn at the end of the lane and passed us on her return run.

I stealthily approached the front garden and peeped over the gate. Breeze was sandwiched between Emma and Jester, eyes closed, neck stretched, face sunward. Ann was right, she was indeed sunbathing. Picking up my camera I located Bernard who was in the barn chopping sticks.

'Bernard, little Breeze has a will of her own,' I said excitedly. 'She is lying with the girls in the garden, and

I didn't even know she was outside. I must take a photo, it's a sight to behold.'

'Hang on a second, I'm coming too,' Bernard said.

The camera lens reflected a line of nodding heads, tongues lolling and panting, under the rays of the hot, hot sun.

'Good God, it's boiling out here, Breeze will roast to death. She needs shade!' Bernard stated, concernedly.

'I could put an umbrella behind her,' I suggested.

'Umbrella!' Bernard repeated, before bursting into fits of laughter. 'I really think she would be better off in the back paddock, as the reeds will shield her from the elements. After all, in her wild state she would be reared in similar cover.'

Bernard's statement made good sense, but posed the problem of how I could keep an eye on her in the paddock and use my sewing machine indoors at the same time. After long consideration, I decided that I would reverse my work pattern and do my hand sewing outside during the day, saving my machine work until the evenings.

The sheep contentedly chomping away in the paddock were given the 'bums' rush', I took up the outdoor life, and the girls moved from the Costa Del Front Garden to the Playa Del Paddock. The views were not as good, but the shade of the old oak tree would stop them from roasting.

Having given Breeze her picnic lunch bottle, I nestled her in a clump of rushes, caressed her and pressed her down. Breeze obeyed this, her first instruction. Her schooling had begun.

Emma eventually realised that finding Breeze by nose-diving every patch of reeds was not the name of the game. This was a new game to her, called 'hide' only, not quite the same as the one where you also 'seek'. Somewhat confusing and boring, she thought, as she sighed heavily before finally succumbing to a snooze.

Laying the Foundations

As I sat beneath the old oak, allowing my senses to melt into the nature around me, I thought I could hear the tree growing. The delight of the bird song emanating from the forest, the drone of bees busy harvesting the last of the May blossom on the hedgerows, the gentle flapping of butterflies' delicate wings alighting on a flowering nettle, all enchanted my senses.

I had grown to love the ever-present animal and insect noises, sounds that for so many years as a city dweller I didn't hear. Visiting friends' remarks such as, 'I couldn't live out here, it is so quiet,' never fail to bring a smile to my face. The countryside often referred to as 'peaceful and quiet' is a bit of a fallacy. It most certainly is not quiet, well, not to the discerning ear, that is.

I watched the pigeons, flying in ever-increasing circles over the meadow sprinkled with buttercups and daisies. My eyes travelled with them beyond the meadow and towards the clearly defined Lakeland hills. The visibility was truly excellent today. The sounds continued with the sheep tearing up delicious mouthfuls of new grass. One coughed, sounding like a pair of deflating bagpipes. Essie's parleying peacocks challenged the air, rising high above the cuckoo's two familiar notes.

Alas, the 'silence' was shattered by the ringing of the alarm clock. It was time for Breeze's lunch.

My high-pitched call of 'Where is Breeze?' was instantly answered as she emerged from her resting place, and covered the distance to me in one great graceful bound. She guzzled down her bottle and returned to the concealing rushes to await her three o'clock feed.

The sun now began filtering through the upper foliage, burning the back of my neck. It was time to move my chair into deeper shade, but I had difficulty finding a flat spot. The normally soft mounds of earth created by the moles looked like mini volcanoes, dry and crusted. This long spell of unusually hot dry weather had no doubt sent the moles deep down into the earth in search of worms.

I resumed my sewing, pausing to re-thread my needle.

'Snap!' My concentration was broken by the sound of Emma, snatching at the air with lightning speed. The fly hovering around her eyes had a narrow escape. How Emma hated flies. She continued her afternoon nap.

Bernard's sudden appearance in his going-out clothes startled me. 'I've made Breeze's bottle, as it's almost three o'clock, and a nice cup of tea for you. The Beehive craft shop rang from Perthshire with an urgent order for hedgehogs, cushions, lavender bags, toilet-roll covers and pot-pourri, which I've packaged up and will post. We are almost out of hen corn and dog food, and I have some bills to pay, so I will go into Carlisle. Is there anything that we're short of apart from coffee?'

'Honey,' I replied, 'as I seem to be living on it these days.'

Bernard kissed my head tenderly. 'See you at about six o'clock,' and quietly slipped away.

Breeze was more than ready for her mid-afternoon milk. Her vigorous sucking almost pulled the bottle from my hand. Her strength certainly belied her size. In no time at all it was empty, then rubbing the teat up the side of her cheek and pulling at it again, she enquired innocently for more.

'All gone, Breeze, there isn't any more,' I replied guiltily. With total acceptance of the 'tap' having been turned off she meekly trotted back to the enveloping reeds to await her next call. She was proving to be as bright as a button. I was immensely proud of her, my star pupil.

I threaded my needle and took up my sewing once more. I envied the girls basking in the sunshine, all sleeping and snoring gently. Oh, what I would give at

Laying the Foundations

this moment to be able to join them. The heat of the day matched the pressure inside my head as I continued my round-the-clock surveillance.

Knowing I was a magnet to midges, and that they would converge on me as the sun began to go down, I began to gather up my sewing. My sudden rush of activity awoke the girls. After much yawning and head shaking, they arose and stretched their legs. It was almost six o'clock, time to prepare their evening meal and Breeze's too.

Our timing was perfect. Bernard was just driving in.

'Where's Miss Peeps then?'

'I've left her in the reeds, I was going to warm her milk before I called her, Bernard.'

'I'll do it whilst you mix the girls' food.' Bernard's offer was greatly appreciated by Jester in particular, whose sigh left us in no doubt that she was deeply affronted at the thought of a deer taking precedence over a great dane.

The girls, having partaken of their dinner, retired to the lounge, with the exception of Emma. She followed me outside upon hearing me shout, 'Where's Breeze?'

She was eagerly awaiting her call and emptied her bottle in double-quick time. Then, responding to my instruction of 'This way, Breeze', she obediently followed Emma and me back to the house.

With the chuckies, pigeons and rabbits fed, it was time to think of our own tummies. My cooking was negligible these days, since we had little interest in food. In fact, we seemed to have been living on a diet of anxiety this past week, and we both now felt faint from tiredness and hunger. We decided on a plain omelette with a side salad.

Emma had decided to do the baby-sitting in the utility room, which afforded us the luxury of watching a wildlife programme on TV while we ate. We washed up, then were enjoying a cup of coffee, discussing the evening's craft work ahead, when the sound of 'peeping' intruded into our conversation.

The baby-sitter appeared with her charge in tow, emitting 'peeps'. Emma's expression of apology, coupled with an enormous sigh, said, 'I can't get her to stop "peeping", Daddy, I think she must be hungry.'

Breeze trotted up to me and licked her lips. There was no doubt about it, she was asking for food.

'It's only half-past seven, Breeze. At this rate it will be cheaper to get a goat,' I replied irritably. Our daily 'pinta' had become a daily litre and was fast approaching a daily gallon.

Her hunger having been sated, she left the kitchen then, pausing at the lounge door, entered and automatically curled up beside Emma on her mattress. As we watched Baba lovingly lick Breeze's head, breathing long sighs of happiness, it was evident that she delighted in Breeze's company. I left the loving duo and disappeared upstairs to my sewing room whilst Bernard remained in the lounge stuffing a pile of cushions with fluffy flock filling.

Breeze's next feed coincided with the girls' supper time, before which they were taken into the paddock to allow them to empty their bladders, 'Pee-pees, girls,' being the cue to make their exit. I pulled on my wellies, by which time they were all waiting in line, including little Breeze.

Night had fallen when I opened the door, and Breeze stood as if glued to the door-step. Her neck seemed to grow by the second as her head darted back and forth, continually searching the blackness. 'Us little deers are frightened of the dark,' she 'peeped'.

Emma stopped in her tracks, realising Breeze's reluctance to venture forth. Retracing her steps, she faced Breeze and grunted, after which Breeze confidently ventured forth. The flow of communication between animals is beyond human understanding, I thought,

Laying the Foundations

as I switched on the outside light and led them into the paddock. Having answered her own call of nature, Emma was not delinquent in her duty toward Breeze. She licked her genital area, stimulating the excretion of urine, just as she would one of her own puppies.

As usual, I had to wait for Jester. She was always the last to finish because she would stand motionless for minutes, forgetting what she had come out for. Even after being prompted to get her act together, she would dawdle about for ages. As a result, many is the time in gale-force winds or torrential rain – or both – that I've had to shelter in the out-house next to the paddock to avoid catching my death of cold.

Sheba and Emma took leave of the paddock, knowing full well I would be there for some time. As Breeze began her 'Feed me, feed me' song, I returned indoors with them, leaving Jester to do her mooching.

Breeze, having polished off her bottle, sought the company of Emma again. Completely oblivious to the fact that she was enjoying her supper, she plonked herself down on Baba's biscuit pile.

Emma's 'Excuse me, I would like my biscuits' plea, went unheeded. Something suddenly caught Breeze's eye, and up she jumped quickly. It was Jester returning. Emma seized her opportunity and gobbled up her biscuits in double-quick time, just as Breeze sank back down again.

Emma gave her baby a night-night wash whilst Jester, the last diner of the evening, finished off her biscuits.

The noise of licks and crunches gave way to gentle snoring as we vacated the lounge and wearily bedded ourselves down in the utility room. My eyes could barely focus as I entered up my diary, the last line of which read, 'The seeds have been sown . . .'

My diary entry on the following evening read, 'Day nine, Breeze blossoms as we wilt!'

9
Flora and Fauna

Our morning programme on awakening had always been a full one, beginning with dogs out, hens out, pigeons to feed and humans to breakfast in this order. It now had an extra item, that of Breeze's performance. Her dance around the kitchen 'peeping', rubbing herself up and down my leg, clamouring to be fed, had qualified her for top billing and relegated us to the chorus line. She couldn't understand why I was unable to produce her breakfast upon demand.

I reminded her that if she insisted on freshly drawn milk, she must wait until the goats had obliged before I could go and collect it. Despite having increased her amount, the jumbo-sized bottle was polished off in double-quick time, followed by her portrayal of Oliver Twist. Her consumption seemed to be insatiable.

As the days hurtled by, our walks were again an exercise in togetherness, and Breeze's voluntary accompaniment was a source of great delight.

Normally fawns would not accompany their mother for the first four weeks after birth, yet here she was, hardly a fortnight old, delighting us with her eccentric frolics. One moment ambling along, her nostrils twitching at whatever Emma had stopped to sniff, then leaping up in the air like a spring lamb with a suddenness that took Emma by surprise, Breeze would tear off with incredible speed as if in training for the hundred-metre sprint.

Jester, Sheba and Emma, convinced that the signal had been given for the race, joined in. As Breeze rushed wildly around the field with reckless abandon, darting in and out of the reeds, the girls tried to keep pace, her swift changes of direction leading them all a merry dance. Her rapturous little face as she circled around and around seemed to say, 'Oh, what a fun game!' She obviously adored being chased.

When the circuit had been completed umpteen times and giddiness had overtaken the girls, Breeze would return, her rapid chest movements allowing her to draw breath only in short sharp gasps. Beaming happily, she rubbed her face affectionately against my leg. A roe fawn must surely be one of the most beautiful gifts of nature, I thought.

Breeze's routine was now established. Her days were spent in the reeds between feeds and walks, and her evenings sharing Emma's bed. Another waterproof cover had been purchased and placed over Emma's mattress, along with a double layer of towels and nappies and finally Emma's fur fabric cover. *All* had to be washed daily, and so a second washing line was hung up. Although Breeze was inodorous herself, her pee could only be described as Eau de Putrescence.

As I finished pegging the last of Stinkpot's washing on the line, I thought I heard the rumble of distant thunder, yet not a single cloud interrupted the blue sky. Having decided it must have been a passing aircraft, I returned my washing basket to the utility room and picked up my sewing materials, reading glasses, straw hat and folding chair. I was all ready for my afternoon's work.

Flora and Fauna

As I closed the door behind me and walked towards the paddock, a thunderous roar filled the air. Charlie had arrived with his harem for his long summer holiday. I could see his two tons of burly brawn parading the field as he bellowed the fact that he truly was a magnificent bull.

Normally I was overjoyed to see him, but today I bemoaned the fact that I had yet another problem to contend with: the possibility of Breeze being crushed under hoof as she lay concealed in the undergrowth. I resolved to keep a watchful eye out for Charlie, and ensure he didn't come too close for comfort.

In the natural order of things, the day went without incident, but the evening had yet to begin.

All thoughts of a leisurely after-dinner amble dissipated as we set foot on the lane, Breeze being in a madcap mood. The girls were driven to distraction trying to keep up with her successive circular bounds, and when she finally came to an abrupt halt in a clump of buttercups, I turned away, envisaging a mass canine collision.

'Breeze is eating buttercups!' Bernard's exultant exclamation had me spinning around like a whirling dervish. Beheading another and another, the glossy golden cups were gobbled with gusto. Gathering bunches as later offerings for the novice herbivore, we skipped home. Whether it was the buttercups, or just part of her growing process, Breeze became hyperactive that evening. It began with the bones. Two or three times a week, depending on the generosity of Peter the butcher, the girls were treated to marrow bones which, with our other preoccupations, were by now long overdue. As they had been paragons of virtue over the last few days, we decided to show our appreciation. When Bernard called from the kitchen, 'What's Daddy got for his girls?', they all rushed in. After proudly accepting their reward, they all returned to their respective beds for a long gnaw.

'Would you like a cup of coffee before I disappear up the stairs to sew?' I asked.

'I'd love one. I'll just pop out and put the chuckies away meanwhile.'

It was a full ten minutes before Bernard returned. He was not alone.

'We have visitors, pet, or should I say, Breeze has,' said Bernard.

Amy and Eddie had called to see the foundling. They had kept in touch by telephone, enquiring as to Breeze's welfare, and were naturally over the moon to hear of her progress. Bernard motioned silently towards the lounge, not wishing to alarm the sleeping beauty. As they entered on tip-toe, their eyes and ours widened with bemused wonder.

Breeze was not sleeping as expected, she was tugging on Emma's bone. The sight of a delicate fawn wrestling with an enormous marrow bone was extremely comical.

I was amazed at Emma's calm acceptance of the situation, knowing her normal reaction if any of the girls so much as looked at her bone. Yet here she was allowing Breeze to take it from her mouth, literally. Breeze continued to hold her audience spellbound, and Emma remained patience personified.

Breeze's visitors eventually departed. Glancing at the racing kitchen clock, I went up to the sewing room. As I set aside my quota of cushion covers for the evening, I felt instantly intimidated. I took a deep breath and started up my sewing machine. Two used cotton reels later, I checked on my progress, and the realisation that only eight were still to be sewn up sent a sudden surge of relief right through me. Finally the pile of cushions that had seemed so everlasting was finished. The clock struck the witching hour as I returned downstairs to the lounge to collapse.

Bernard was just in the act of giving the girls their

biscuit snack. Emma was in one of her turned-up-nose moods. 'Don't want dog biscuits tonight, I fancy custard creams.'

Naturally Fussykin's request was complied with and placed in front of her.

Breeze's sudden leap from sleep to liveliness startled us both. She sniffed curiously at the custard creams then, with unflinching certainty, snatched one from beneath Emma's nose, rattled it around in her mouth, crunched it then swallowed.

I watched Emma expectantly, waiting for her to snap as Breeze stuffed her mouth full with yet another, but Emma's willingness to share whatever she had with Breeze exhibited true love indeed as they competed with each other for the last remaining biscuit.

Emma, gracious in defeat, then moved her body along to allow Breeze space. She sedately lowered herself, tucked both front legs under her, snuggled into Emma's body, and gently lowered her long lashes in slumber. Baba's expression, which can only be described as doting, was touching to see as she turned her head on one side and sighed with happiness before closing her eyes.

I yawned, I was battling to keep awake. Bernard appeared with two mugs of cocoa.

'Oh, you look done in, pet,' he said softly. 'Why don't you sleep in our bed tonight? I can see to Breeze. I think actually that we should lengthen the time in between feeds to four hours, as it may encourage her to forage.'

I agreed and felt a little less guilty knowing that Bernard, who looked drained and exhausted himself, would at least be able to sleep until four-thirty in the morning. As we were on the go by seven, and I was bringing home the milk at seven-thirty, it would seem to fit in very well.

I drank my cocoa, bade my family 'Night, night', and climbed the stairs. Oh, how soft the bed felt, as my head sank into the pillow, softer than ever before. I felt a seeping fatigue overtake me as I drifted off into sleep and the curtain came down on yet another day. For me anyway.

I wasn't to know that Breeze was to take a curtain call.

I awoke before my early-morning alarm call to the sunlight flickering through a chink in the curtains and the smell of toast wafting up the stairs. How strange it felt to wake up in bed alone, but how wonderful this feeling of clarity. I skipped down the stairs, three at a time, and opened the lounge door. Fighting my way through dogs rushing at me in greeting and a fawn clamouring for her breakfast, I was reminded of sale time at Harrods. It was quite an achievement to reach Bernard in the kitchen in one piece.

'Good morning, pet, did you sleep well?' Bernard asked as he kissed me on the nose and handed me a cup of tea.

'Blissfully, thank you, I feel ten years younger,' I replied with a smile. 'What kind of a night did my loved one have?'

'Are you referring to me or Breeze?' Bernard answered, his mouth in an inquiring smile.

'You of course,' I responded politely, but puzzled.

Bernard sighed. 'To say I've had a dreadful night would be putting it mildly, but mischief over there,' Bernard paused pointing his finger emphatically in Breeze's direction, 'had a rare old time of it!'

Bernard's tone told me that Breeze had blotted her copybook. 'Oh dear, what happened?' I asked sympathetically.

'What *didn't* happen is more like it,' he replied irritably. 'Firstly, Breeze's alarm clock did not coincide with the alarm call, she was demanding to be fed at three-thirty. Having told her firmly that in the wild, food does not come at mealtimes, it comes when it is found, she then went searching. Finding the buttercups, she decapitated them, knocked the jar over, glass and water everywhere, then panicked and scooted into the lounge. As I was mopping up the mess, crash, bang, wallop, squeaks and yelps from the lounge, poor Jester was enmeshed in plant stand, plant pots and soil, Sheba was cringing beneath the rocking chair, there was soil all over the carpet...'

I was having great difficulty in keeping a straight face as Bernard paused for breath, then continued... 'whilst *she*, Angel Face, was cuddled up to Emma looking the picture of innocence. Only the various bits of green stuff sticking around her mouth betrayed her guilt. Anyway, I spent the next half hour cleaning up the mess, then put the kettle on, which was *her* cue to squat and pee all over the carpet. Can't you smell the disinfectant? I think I used the whole bottle. Consequently, I haven't slept and feel bloody awful!'

I was fighting hard to keep my composure and not break into hysterics, as I spoke. 'Poor darling, when I've collected Breeze's milk, why don't you go up to bed for a few hours?'

'Thanks love, it's either bed or emigration,' Bernard yawned, 'I've been on automatic pilot all night.'

Breeze, believing that her continued 'peeps' and leg rubbing had gone unnoticed, was now preparing to have a grand mal to attract my attention, so cramming a slice of toast in my mouth and taking a quick gulp of coffee, I departed.

Later, a cloud glided past the sun, blotting it out as I settled down in the paddock with my sewing. The girls were by my side, Breeze was nestling in the reeds, and Bernard was in the land of nod. An air of rustic peace prevailed. But not for long! By mid-morning the clouds had gathered, and the still, hot air was invaded by squadrons of midges. Their mission, to search for moisture. Having found their target they dive-bombed the girls, attacking their eyes, nostrils and mouths.

I led the irritable, walking wounded to the safety of the front garden. The evacuees sighed with relief.

We had gravelled this area and placed pots of flowers on it when we moved in, and this had proved to be a prudent decision, as the midges much prefer grass in which to hide. The day was extremely sultry, the kind of weather that saps one's energy. Even the birds were singing drowsily whilst the girls looked comatose.

I stood up and stretched my arms above my head to ease out a crick in my neck. A sudden roar made me jump, then shrill peeps developing terror-stricken notes had me racing out to the paddock as fast as my little legs would carry me. Breeze, fear mirrored in her eyes, was flying around dodging this way and that in an effort to escape a confrontation with Charlie, the bull, who had obviously disturbed her as she lay concealed in the reeds. He roared again, snorted and tossed his head.

I felt a surge like an electric current pass through me as I temporarily froze, my heart thudding. Throwing caution to the wind I took a deep breath and charged forward, waving my arms wildly in the air, screaming 'Huf-huf!' at the top of my voice. Charlie stood his ground stubbornly. My irrational heart urged me to gather Breeze up in my arms, turn tail and run, but my rational head told me this most certainly was not the thing to do. I felt powerless to do anything other than to force myself to slow my pace and retreat backwards.

Feeling like a jelly that hadn't quite set, with my arm outstretched holding an imaginary bottle, I called 'Breeze, where's Breeze?' in a quivering voice.

Breeze's trust overwhelmed me as she unhesitatingly responded to my call and followed. My fingers trembled as I fumbled with the gate catch. The relief in having reached the farmyard made me feel quite lightheaded. I closed the gate firmly on Charlie. We had survived our ordeal.

Flora and Fauna

Emma's eyes lit up as Breeze ran to her and burrowed herself into her body. Safe in the knowledge that her family would protect her, Breeze closed her eyes in exhaustion.

All the commotion had naturally awoken Bernard and put paid to his day of rest. He had gone to bed with the threat of a headache and had now developed a full-blown migraine. The pain was evident in his eyes as he beat his fist against his brow. My own mind, having been whirled into panic by Charlie, had tensed me up, and a familiar thumping was settling into my head. It was cocktail time!

I handed Bernard his glass of bubbles and he raised it to mine saying 'Cheers!' After drinking it rather than gulping it, I wrinkled my nose in distaste. Not the most pleasant of tablets to take, still, they were very effective.

Bernard yawned and stated, 'I feel like a zombie!'

'I'm not surprised, pet, painkillers and no sleep are a potent combination. I'm somewhat lethargic myself, and I had a good night's sleep,' I replied.

As neither of us was capable of concentrating, craft work was out of the question. We opted to do some gardening instead. The eight trays of bedding plants, courtesy of Chris and Jean, had lain untouched for over a week and were crying out to be permanently housed, almost in full bloom.

By the time the final plant had been placed in the flower tubs, we were close to wilting. All that remained now was to water them all in. Bernard and I took our leave from the garden to have a coffee break. Refreshed, we returned with our filled watering cans. Wordlessly, we stared in astonishment. Darling Breeze was delicately decapitating the flowers we had just spent an hour in planting. She then sauntered over to my cherished rose bush, sampled its petals, then began stuffing her mouth full with its beautiful blooms.

My natural ebullience vanished instantaneously.

BREEZE: WAIF OF THE WILD

'Stop that this minute, Breeze,' I shouted in a scolding tone. Resisting my command, she carried on cramming in the roses. 'No, no, naughty girl!' I screamed, darting forward.

Bernard took hold of my arm pulling me back. 'You'll frighten her, pet, it's pointless chastising her, she doesn't understand. To her it's all just nice tasting munch, and you should be overjoyed at the fact that she is now foraging, albeit at the garden's expense.'

He was quite right, of course, but I was not in the mood. All I could think of were the repercussions of Breeze's drastic pruning. My highly prized rose bush, a Christmas gift from my dear friend Daphne – which had been voted rose of the year, no less – was about to be stripped bare. I had been so looking forward to including its fragrant petals into my pot-pourri.

What pot-pourri! The thought suddenly struck me that there would be no flowers at all to harvest if Breeze continued her rampage. What about my mother's hobby of pressed flower pictures? She relied on our garden to keep her flower-press full, spending the best part of her annual holiday with us, selecting, gathering and pressing them. The artistically framed pictures she so enjoys making help shorten the long winter nights, and are always well received by family and friends at Christmas.

She would be visiting us soon, so to avoid her packing her flower-press to no avail, I determined to drop her a line. 'Sorry, Mum, no flowers to press this year. Breeze, the little deer, ate them all!'

'Having nurtured my garden for months and now being forced to stand by and watch it disappear before my very eyes takes some doing, Bernard,' I said, self-pityingly.

'Just think of Breeze as your own personal gardener, pruning and weeding for you.' Bernard, making light of the situation, brought a smile to my face.

Having regained my composure, I resolved to turn a blind eye to Breeze, the garden gourmet. I picked up the watering can and went diligently about my watering in of the headless bedding plants.

The evening drew to a close, as did our eyes. It was evident that the past days had exacted their toll on us both.

We put the bunches of selected grasses we'd picked on our early evening walk into two buckets and placed them on the floor for the midnight marauder. Satisfied that all were settled down for the night, we quietly closed the lounge door behind us and *both* staggered wearily up the stairs.

10
The Sweet Smell of Success

I shot into wakefulness at the sound of the alarm clock heralding the new day. The moment I opened the lounge door it was apparent that something was wrong. Instead of the girls' exuberant greetings, all was quiet. I felt instantly apprehensive as I pulled back the curtains allowing the sunshine to fill the room. My concern was heightened to find Emma sharing Jester's bed, Sheba under the settee and no sign of Breeze. Their expression needed no translation, someone had done 'a naughty'.

Looking around for the evidence, my eyes alighted on some black, shiny, oval objects about the size of a fingernail, resembling currants. These were interspersed with everlasting flower heads and fir cones, what now remained of my recycled dried flower arrangement. Poor Emma, little wonder she had taken refuge with Jester. Her own bed was covered in the same currant-like things that adorned the carpet. They looked curiously like deer droppings. A sweet earthy smell now invaded my nostrils and confirmed my suspicions.

I sighed despairingly. 'Where is the little monster?' Emma sidled up to me with a look of acute embarrassment and indicated the utility room. Her reluctance to accompany me immediately set alarm bells ringing.

So much for my anxiety! My eyes almost popped out of my head. I stood dumb with amazement. The room looked like a pigsty. My vegetable box had been emptied: potatoes and carrots, with bites out of them were rolling around the floor, and all that was left of my cabbage was a few stalks. The leeks at least remained uneaten, although they too had been sampled. I made a mental note that 'deer find leeks unpalatable'.

It was obvious too that our choice of grass was not to her liking as it lay untouched, strewn across the floor.

Amidst the chaos of the room, Guiltless Greedy Guts lay in an embryonic position, dead to the world. Curbing the reprimand on my tongue, I picked up the dustpan and handbrush and tip-toed out.

As I crawled on my hands and knees across the lounge carpet cleaning up the party leftovers, Emma volunteered her help. Sniffing and studying the carpet, she stopped at each pile, inclined her head to one side, wagged her tail and announced, 'I've found another one for you.' I was thankful deer didn't eliminate like cows; at least these brushed up easily enough and the smell was bearable.

Although I wasn't one to baulk at cleaning up after animals, having helped raise four dogs from puppyhood, it is still, nevertheless, an affront to the senses, and not all that good for the carpet either. The upstairs toilet flushed. This was normally the cue for the girls to vie excitedly for first place to greet their Daddy. This morning they lay motionless.

As Bernard opened the door, a look of incredulity passed over his face. 'Oh . . . God! I see Breeze has been up to her party tricks again. What's that?' he asked, looking pointedly at the shovel.

'Breeze's organic contribution to Coldside's compost

The Sweet Smell of Success

heap,' I replied peevishly, as he scrutinised the shovel's contents.

'Oh, what a perfect specimen!' he declared, almost bursting with pride. 'It smells quite sweet, doesn't it?'

I managed a weak smile, pressed my forefinger against my lips in the 'keep quiet' gesture, and beckoned Bernard to the utility room. His sweeping glance took in every detail of the room before alighting on the deceptively angelic Breeze. His reaction was simple, he burst into laughter.

'I'm glad you think it's funny, she is a little devil,' I retorted.

'Oh, don't be cross, love, just look at her. Any moment I'm expecting her to sprout wings and fly,' he replied, as his laughter rang out. Bernard's laugh, so infectious, broke the bounds of my restraint, and I joined in.

Unable to resist the happiness that they can sense when we laugh, the girls burst into the room, their eyes shining and tails resembling windscreen wipers on full. Breeze yawned lazily, lifted her left hind leg and scratched behind her ear, just like a dog, then licked her lips in anticipation of her breakfast. As she uttered her first 'peep' of the day, I hurriedly set off on my early morning milk round, Bernard's laugh following me all the way.

Over the next few days we continued to dance to Breeze's tune. She now performed all three functions of sleeping, eating and eliminating, to perfection, the latter being in different places. As she had no control over her excretory activities, it was useless to reprimand her. Emma didn't understand why Breeze shouldn't obey the rules of the household, constantly complaining and justifiably so: 'She has eliminated on my sponge again, can't you lay papers down like you did with me, or immediately whisk her outside?' As these training methods were out, we had to adopt the eagle-eyed approach. Fortunately she did follow a preliminary ritual, like sniffing around for a bit or making scrapes on the carpet, which enabled us to whisk a towel under her. As she mostly parted with her droppings 'on the move', the sight of me crawling on all fours, pushing towels in her direction, was a bit embarrassing. We had a problem that couldn't be solved overnight, but something would have to be done, and fairly quickly, as the bills for disinfectant in an effort to obliterate all traces of smell, were mounting at such a rate that shortly it would be less expensive to replace the carpet.

By now Breeze had become familiar with the topography of the house. After the upstairs had been discovered and 'christened', it was immediately declared off limits by barricading the open staircase off with a clothes airer. As hectic washday followed hectic washday, the washing machine finally committed suicide due to overwork, and the utility-room carpet was cremated.

As the whole house became a toilet facility, we evaluated the situation, and planned our next course of action. One of the out-houses was now to become Breeze's bedroom. We would take her in last thing at night as we retired, and let her out in the morning. Nothing else would change, and we would still continue to gather grass and place it in her sleeping quarters. I felt quite relieved at having found a solution to the washing problem, and also that we would be able to go to bed, safe in the knowledge that the house would be in one piece when we awoke.

Breeze's very own stone-built bedroom, complete with door and window, was twelve metres in area, well lit with fluorescent tube lighting, and only metres away from the house. Golden straw formed the carpet, a mound of freshly mown hay was against one wall and her very own bed – the base of an old settee – against the other. Armfuls of selected succulent grasses collected by Bernard were spilling out of a feeding trough. A nice

The Sweet Smell of Success

juicy lettuce, the outer leaves of a freshly dug-up cabbage and a container full of dog biscuits for 'afters' (having omitted the diamond-shaped pink ones that she wasn't too keen on), completed the buffet.

'Do you think Breeze will find it to her liking?' Bernard enquired.

'Like it . . . she will love it,' I enthused. 'It's fit for a princess.'

Over tea, we debated how best to persuade Breeze into the royal room, and decided the best course of action would be to await her next feed at around eight in the evening, entice her in with the bottle, leave her for an hour then let her come back in the house with us. This way, when we had fed and closed Breeze in her bedroom for the night, she would know we were not abandoning her. She would know she would see us again in the morning, and a routine would be established.

Breeze awoke in her usual vociferous, lip-licking, anticipatory mood. The sight of her bottle being prepared had her cavorting around the kitchen like a cat on a hot tin roof.

'Does Breeze want her lovely din-dins?' I asked temptingly, thrusting the bottle toward her.

As she lunged forward, I quickly withdrew it, did an about turn, and headed in the direction of the out-house. She needed no encouragement. With her eyes fixed resolutely on the teat, she chased after it, totally oblivious to the fact that we had now entered her new bedroom. I closed the door behind me as her mouth latched on to the bottle.

Her long-lashed eyes remained half closed whilst she contentedly sucked on the teat, then having opened fully at the realisation that the bottle was empty, dilated as she took in her unfamiliar surroundings. Her eyes gazed back miserably. I felt utterly deflated. Taking a deep breath, I hurried out, her 'peeps' of protest at her desertion sounding in my ears.

Bernard was about to water the flower beds – well, what was left of them. He took one look at my dejected face, laid down the hose-pipe and enquired, 'What's the matter, wouldn't she go in with you?'

'Oh, she went in okay, but she doesn't like it.'

'Well, she will have to learn to like it,' Bernard replied, his voice betraying his disappointment. 'She mustn't be allowed to dictate to us. Goodness, she's only the size of tuppence.'

'But listen to her, pet, she sounds so forsaken.' My voice trailed off as a lump came into my throat.

'For heaven's sake, Marie, she has only been in there five minutes. She'll settle down shortly, you'll see,' Bernard assured me with such emphasis, I had to agree.

Turning my attention to unravelling a kink in the hose-pipe, I tried to block out Breeze's distress calls, but minutes later her relentless pleading, accompanied by loud thudding, had me dissolving into wretchedness.

'I can't stand it any longer, Bernard. I'll have to let her out,' I cried.

'She is staying where she is!' Bernard said emphatically.

'How can you be so callous, can't you hear her launching herself against the door? She's having a nervous breakdown. We'll end up with a dead deer!'

'Oh, don't be so melodramatic. You're allowing her to dominate your life. Well, she isn't going to dominate mine!'

Bernard's unyielding attitude sent me stomping indoors, head bowed in a sulk. Seconds later, my head held erect in defiance and jaws clenched, I petulantly marched outside.

'You are breaking my heart,' I snapped accusingly. 'I'm spending the night with Breeze, in the out-house, and tomorrow,' threatening him with the ultimate, 'I'm leaving home!' I stormed away.

On opening the door of the out-house, the sight of

The Sweet Smell of Success

Breeze, fraught with confusion, swamped me with compassion.

'Mammy's here, my darling, she will stay with you,' I whispered in a choked voice, cradling Breeze like a baby in my arms. My words of assurance brought forth a 'peep' that was unmistakably triumphant.

I switched my gaze to the door as it opened cautiously, and a head peeped in. Bernard's expression was one of concession.

'Oh, what a lovely room, fit for a princess, I think Mammy said.' His impish smile revealed his sense of humour. 'I'm sorry, pet, I was too absorbed in what I had perceived to be the solution to our problem, I hadn't taken Breeze's reaction to the situation into account,' he said, kissing me gently on the cheek. 'And I know you'll work something out.'

I was too touched to answer. Bernard's faith in my ability to solve almost any problem made me feel quite virtuous.

'How about a nice cup of your coffee?' Bernard's request had hardly left his lips when Emma, Jester and Sheba appeared. Their request came higher than Bernard's need, so all six of us walked down the lane in the late evening sunshine.

11
Wet, Wet, Wet

I don't think the sun went to bed that night judging by the heat of the following morning. The kitchen sounded like a railway station in the days of steam engines, as the girls huffed and puffed irritably.

'I'll tell you what,' Bernard said, 'let's have our walk by the river. The girls can have a splodge and cool off.'

Understanding, as they do, every word we say, they chorused ecstatically, the operative word being *cool*. We crossed the lane in front of the house and began our downhill saunter across the open field.

'The sun is boring a hole in my head,' I complained, as we wended our way through sheep and lambs, Indian file, Breeze bringing up the rear. My heart went out to the mobile woolly jumpers. Whilst their thermal coats are ideal for wintering on the hills, today they must be insufferably hot. It was no good them protesting, they had to grin and bear it. The farming timetable doesn't make allowances for an exceptionally hot early summer, everything follows in its time, and there was a whole month to wait for shearing. How lucky I was not to have been born a farmer. The only way I wanted to work with animals was to conserve them.

I couldn't help feeling guilty as I took off my blouse and tied it around my head, Arab-style.

'There now, this will stop *me* getting sunstroke... sorry, little woollies.'

Bernard's suggestion had seemed like a good idea at the time, but was losing its appeal by the minute. I was far too hot and animal sensitive to enjoy our walk.

'Hang on a minute, Bernard, I'm melting.' I peeled off my tee-shirt exposing my summer vest.

'Good God, Marie, I'm not surprised you're sweltering, you have more layers on you than an onion!'

Bernard's statement was never truer. I *was* guilty of overdressing, and had been since moving here. The temperature difference from south to north had hit me noticeably, and I still wasn't acclimatised. I thought of how many remarks from friends seeing me in the summer of, 'Gosh, you have lost weight,' were due to my having shed a few skins.

We finally reached the meadow. The closed gate indicated Mitch's wish to keep the sheep with single lambs separate from those we had passed with twins. Baas and bleats ensued as we opened the gate and entered the meadow. The sheep having called their respective lambs to 'heel' all moved up to the far end of the meadow abutting George and Essie's farm.

The sight of the water – or was it the smell? – saw the girls break into a trot, putting me in mind of a string of camels having found an oasis in which to top up their tanks, as they gulped down the cooling water.

'I don't think I've ever seen the river so low, have you?' I turned to address Bernard, to find him preoccupied, shielding his head with his hand and looking beyond a clump of gorse, where a lone lamb stood bleating forlornly.

'Keep the girls with you, I think there is a ewe cowped.' Hurriedly he went to investigate.

Us 'townies' had been educated in the usage of

farming terms by Mitch. 'Cowped' means a sheep is down, lying on its back and unable to get up, we were informed shortly after befriending Mitch. At this time of year, being heavy in fleece, they have a tendency to cowp and, once down, find it impossible to rise – hardly surprising when you consider the weight of a fleece. Mitch had also said that if they were not helped to their feet within two hours, the result can be fatal, due to the pressure on their lungs. This thought made me feel sick in the pit of my stomach. Suppose Bernard was too late? I shuddered.

Watching him grasp the ewe by its horns and rock it back and forth, my spirits soared. It was obviously still alive, otherwise he wouldn't be doing this. As he continued rocking, more briskly now, gathering momentum, the sheep was returned to its feet once again. She staggered for a few moments, then lurched forward determinedly in the direction of the lone, bleating lamb, calling fervently. Maternal instinct must be one of the strongest of emotions, I thought. Here she was, close to death's door, and thinking only of her offspring.

Two simultaneous sounds broke into my musings. Bernard's voice, 'Where is Breeze?', and the sound of thrashing water.

The river bed falls away into deep pools and Breeze in her naïvety had followed Emma and was now out of her depth. My first thought was that I was looking at Breeze swimming. It took me a moment to realise that she wasn't swimming at all but was in trouble, chest deep in water, and struggling desperately to find solid ground. Her gestures of helplessness set my pulse racing. I plunged into the river and gathered her in my arms. As I carried my little wet bundle to the safety of the river bank, a great wave of remorse washed over me.

'I told you to keep an eye on her!' Bernard's reprimand

was the last thing I needed. Unable to stifle my emotions, the child in me took over, and I began to cry.

Cupping Breeze's face in my shaking hands, I gazed fixedly into her velvet brown eyes searching for the appropriate words with which to apologise. I could only manage a gulp, tears choking my power of speech.

Bernard's mood softened at once and his warm arms of comfort enfolded me. 'There now, there now, I'm sorry, pet.' My tears never failed to bring out the chivalry in Bernard. 'I didn't mean to upset you, it wasn't your fault.'

My eyes brimmed with tears again, their full load not yet shed. 'But it *was* my fault, Bernard. Just because she is a wild animal doesn't mean I should leave her to her own devices. When you accept responsibility for a puppy, you don't open the door and say there's the world, go and explore it, do you? Well, just like a puppy, Breeze was born without fear and it's up to us to protect her, guide her, warn her of danger and think only of her future, and she nearly didn't have one . . .' I further sobbed, ' . . . because I neglected my duty.'

'Goodness me, it's not the end of the world, you're making a mountain out of a molehill! Do you remember Jester's first encounter with water? You didn't hold yourself responsible then, did you?'

My mind flashed back to a much younger great dane whose appetite for calamities almost made us change her name to Jane. One minute Jester was leaping blindly through the bracken on that sunny Sunday's afternoon walk, the next she was somersaulting through the air, and landing in a slime-covered pond. Her fright was so great she became immobile and had to be fished out. It had turned Jester off water for life. Now, if her tummy so much as touches the water, she turns tail, and like a big daft lass, runs for dry land. The recollection brought a faint smile to my face.

'That's better, Marie, think constructively, not

destructively. Look upon this incident as advantageous to Breeze. She will now respect water.'

As we headed the wagon train home, even our wellies were melting. The full glare of the sun made the short walk uphill feel like a million miles. I wondered if Charlie had vacated the paddock yet; the girls' usual spot would be suicidal in this heat.

The thought had hardly formed in my mind when Bernard, as if psychic, said, 'There's no way the girls can lie out in the front garden today. Let's hope Charlie has gone so they can sit under the tree.'

Our luck was in, Charlie was nowhere in sight. He and his herd had moved to pastures new, leaving their evidence behind. Picking their way through flattened molehills and cowpats, the girls flopped down beneath the oak tree, thankful for its shady branches.

Breeze was wondering what clover tasted like when Bernard's call of 'Come and get it' decidedly changed her mind.

'What on earth . . . ?' I broke off in a fit of laughter as he held out not one bottle but two. 'She'll never manage to drink all that, Bernard.'

'Want to bet?' he replied challengingly as Breeze, having already emptied the first, was now starting on her second. The speed at which she was able to swallow simply amazed me. It went down quicker than my bath water did after releasing the plug. The second now empty, she began rubbing her cheek in gentle persuasion along the bottle as if coaxing the tap to turn on again.

A broad grin lit up Bernard's face as her sucking mouth sought the teat again. 'You have got to be joking, Breeze! It's all gone now, you greedy little piggy.'

There was more than a trace of indignation in the 'peep' that followed. Bernard exploded into laughter and, shaking his head from side to side, went back to the house.

Breeze's interest in the clover patch was renewed,

and was to her liking, judging by the rapidity with which the clover heads were disappearing. Only when every last one had entered her tummy did she sink down on her haunches next to Emma. She stuck her long spindly legs straight out in front of her, lowered her head until her chin was resting on the grass and finally dropped her eyelids.

I had to blink hard to stop mine following suit. Pulling a large box of hedgehogs to my side, I sighed: all of them had to be packaged by noon, and taken to the Gretna hotels. Bernard had obliged over the last three weeks, but today a stock-take was due, and this was my department. As it was the last day of June, it was also our 'pay day'. I calculated our takings would be down on the previous month, as our supply-according-to-demand strategy had fallen foul of a deer.

'We'll make up the lost revenue next month,' Bernard, the eternal optimist, had stated. 'By then, Breeze won't be so demanding of our time.'

I wasn't totally convinced, but remained hopeful. And with the anticipated arrival of a new washing machine later that day, I could at least put the escalating laundry problems out of my mind.

I hastily grabbed a hedgehog which, obedient to Sod's Law, jumped out of my fumbling fingers, rolled and promptly came to a halt in a cowpat. I let out a sigh of resignation, shrugged my shoulders and ploughed into the box once more.

As I opened the hen-house door the following morning, the hens squawked in complaint at the earliness of the hour. The only way we could achieve our July production target was to lengthen our working day, so we arose before the cock had begun his crowing.

'It's the early bird that catches the worm, chuckies,' I stated, hoodwinking them into thinking it was more for their benefit than ours, as I shooed them outside.

Breeze, on the other hand, needed no explanation.

'An early breakfast will suit me just fine, thank you very much,' she peeped.

'You will just have to wait, Breeze, there is no way I can call on Mary at the crack of dawn!' I might just as well have spoken to the wall for all the notice she took of me. The pressure inside my head began to build with each staccato peep. 'Breeze . . . stop . . . that . . . squeaking . . .' The words were spaced through clenched teeth.

Breeze, convinced she was about to die of starvation, continued her ceaseless clamouring until the Knight-Errant came to her rescue.

'All right, all right, Breezie, I'm sure Mary won't mind, will she, Mammy?'

Having come off second best again, I left on my pilgrimage.

By noon the air was still and sultry and the sun's refusal to shine had caused Jester to take umbrage. She always held us personally responsible for every sunless day. She took a deep breath, stared at the sky restless with racing clouds, and exhaled slowly. She cast a sidelong look in my direction that was darker than the lowering greyness, and collapsed back on her sunbed.

By mid afternoon a light breeze had sprung up and tempered the heat. The sound of thunder rumbling in the distance accompanied us on our earlier than usual walk, and the dark and menacing sky seemed to be resting on the hills. It was evident that rain was on its way. As the growling rolls of thunder came ever nearer, we took to our heels and made for home. Terrified was not a strong enough word for how Jester felt about storms. At the impact of the first thunderbolt directly overhead, she tore off with a speed that could only be equalled by the flash of lightning that followed. Her stride stretched far beyond that of any hound, more akin to a wild horse, as her limbs ate up the ground, homeward bound.

BREEZE: WAIF OF THE WILD

One thing was certain, Jester would beat the impending rain, applying her brakes only when home. Rain drops were now falling, and Breeze bounced into the air as if on springs when they touched her back. Her first impulse upon landing was to shake off whatever was landing on her, but realising she was fighting a losing battle, tensed her muscles hoping this would frighten the rain away. It didn't. She looked skyward, startled, and began shaking herself again. This was her first encounter with rain and, judging by her reaction, she didn't think much of it.

'It's only water, Breezie, makes the flowers grow for you to munch,' Bernard explained gently.

There was nothing gentle about this spiteful downpour, however. Selecting turbo mode, we all raced for home.

Jester's compulsion to get away from the explosive noise continued indoors, and she ran from room to room in a state of extreme agitation. To help muffle the thunderous sounds, our immediate task was to dash around switching on every conceivable noisy machine prior to covering her with a blanket and cradling her closely until the storm subsided.

'There now, there now,' Bernard whispered soothingly as Jester, shivering uncontrollably, buried her head in his wet lap.

'You're soaking, pet, you ought to get out of those wet clothes,' I said concernedly.

'Oh, don't worry about me, Marie, just dry off the girls, and you get changed.' Bernard's remark was typical. Humans always came after animals in our house.

'Well, at least dry your hair, Bernard.' I threw him a towel.

Breeze needed no drying. Emma, looking decidedly broody, reminded me of a cow giving her new-born calf its 'welcome to the world' wash, as she swamped Breeze in a flood of licks. Blissfully enjoying every moment of being mothered, she arched herself against Emma's tongue, eyes closed, revelling in the attention being showered upon her.

'Isn't that the most beautiful sight you will ever see? Who would believe it?' declared Bernard sentimentally.

'It's priceless, pet, memories fade, but photos live on,' I replied, opening the drawer and taking out my camera. If ever a photo captured the embodiment of soul-mates, this one did. Clearly a relationship had been forged that wouldn't easily be broken. I wasn't so sure I was thinking only of Emma, either.

After an interminable hour of the TV, radio and record player all competing for air-wave space at the same time, during which my head felt like a volcano on the point of eruption, the storm abated. As the last peal of thunder died away and the rain eased to a gentle patter, the sound systems were switched off. Peace and quiet now prevailed.

Carrying the freshly made coffee into the lounge, I heard another rumble, this time not overhead, but underfoot, or so it seemed. A smile slowly crossed Bernard's face as another indelicate sound rattled around the room.

With eyebrows raised, I spun around. 'Oh Emma, was that you?'

Bernard was quick to her defence. 'Not it wasn't, it was Breeze.' I was utterly lost for words. A rarity for me.

I had witnessed many funny scenes in my life, but none so comical as a graceful fawn chewing her cud like a cow. It seemed so undignified somehow, her belching away as if afflicted by a bad case of indigestion. Suddenly my tongue untied itself, and I burst into laughter.

Breeze's body left the floor with her second resonant burp, Emma's tongue still in contact. Poor Baba, her tongue was stretched to its fullest extent – not, I should imagine, a very pleasant experience.

'What a bloomin' racket, Breeze,' I exclaimed as her tummy rumbled and reverberated around the room.

'Oh, don't chastise her, Marie, all ruminants cud, it's part of their digestive process. Anyway, it proves that all her working parts are in order.' Bernard paused, then added proudly, 'She's a big girl now!'

'Did you hear what Daddy said, Breeze?' I asked, pushing a dry towel under her rump and tugging the wet one out from under her. 'You are a big girl now, and big girls do not wet the bed.'

Like water off a duck's back, my words went unheeded. We awoke the next morning to another wet day, in both respects.

12
Growing Pains

The problem in our part of the country is that when it rains, it forgets when to stop. It bucketed down for the next three days without even a blink of sunshine. Endless hours of rain I find depressing enough, without the addition of piles of sodden towelling dog driers, floors patterned with muddy cloven hooves and paws, and Jester's insufferable sighs of petulance. We find rain-soaked walks anything other than enjoyable, but dogs still need exercise, regardless of the weather, and in this respect we never shirked our responsibility.

'See how much we love you,' Bernard stated, as we donned our wellies and rainy-day gear, before squelching over the fields.

Breeze's need to keep a steady supply of food passing into her rumen dictated as well, despite the elements, and with no blueprint to follow as to how long she needed to graze for, we decided to individually accompany her, alternating at four-hourly intervals to coincide with her bottle break.

As the dismal dawn of the fourth day gave way to a glimmer of sunshine, Breeze's daily pattern had been established. Her pursuit of individual blades of grass

BREEZE: WAIF OF THE WILD

lasted about an hour, followed by a short period of restful cudding and digesting, before proceeding with more fastidious grazing.

Jester, now showing distinct signs of sun-withdrawal syndrome, pricked up her ears at the sound of the front door opening. The sight of the sun-beds being carried out meant only one thing, the sun was shining. Her transformation was instant. Beaming like a lighthouse, the congenial side of her nature re-emerged.

The aromas and sounds of summer surrounded us once more. A smell of pine drifted in from the woods and song filled the air, as birds serenaded the sunshine. Just like unfurling flowers, the girls offered their faces to the sun.

Shouts of 'Fetch on, Tess' echoed across the meadow, followed by whistles and another yell of, 'That'll do, Tess!' The voice was unmistakably Mitch's. My eyes focused on a huddled bleating mass making their way up the hill. Shearing time was upon us.

As July wore on, I swallowed deeply each morning before opening the door, knowing what I would see when I entered. The carpet was still being liberally sprinkled, and short of putting a nappy on Breeze I had not found the solution. My usual exclamation of condemnation was now ignored and the word 'naughty' had become meaningless. By the second week in July, I sounded like a well-worn record.

Breeze's spots were fading and her 'periods' had begun.

Our first awareness of Breeze's 'plastic period' was when the bag of porridge oats came crashing down. I almost choked on my toast. The entire contents spilled all over the kitchen floor, yet her interest was not in the oats, but the plastic bag containing them. I had been puzzled that morning at finding a sliced loaf on the floor directly under the bread bin with the odd nibble out of a slice, whilst the plastic wrapper was in shreds.

My first and only thought was that we had a supermouse. I now knew who the culprit was. As this obsession continued, I became more and more fed up. Nothing enveloped in plastic was sacred. The morning I found a consignment of carefully packaged cushions, three dozen in all, scattered about the floor, minus their protective plastic coverings, I secretly wished she was a bee and would just buzz off.

This 'plastic period' thankfully passed, only to be succeeded by her 'straw period', which quite honestly drove me to distraction and back to the 'fizzies'. The sight of my large circular straw mat, painstakingly crafted by an African lady, and a memento of our four years in South Africa, lying in shreds had me screaming 'What did I do to deserve you?' at the top of my voice. I lost count of the times I stopped her nibbling the rush seating on the kitchen chairs, the Ali-Baba basket in the bathroom, the rattan-topped coffee table and the multifarious bric-à-brac which had been lovingly collected over the years. To avoid her eating us out of house and home, all were carried upstairs, safely out of her reach.

No sooner had we recovered from this less than engaging period, than she delivered her pièce-de-résistance. It was a sound of metal mixed with china in the lounge that had alerted me, just as I was about to serve dinner. Knives, forks, plates, condiment set and glasses were all scattered on the floor. The explanation was simple. Our woven straw table mats had caught Breeze's eye, and were now in her mouth. Shaking my head from side to side, my temper frayed. I bellowed the 'No!' command, as I tugged the mat away from her, then reset the table once more. At the suggestion of a twinkle in her eye, I delivered a 'Don't you dare, Breeze' threat with such a tartness in my tongue, I hardly recognised it as my own. Whether mesmerised by my pointing finger or vigorously shaking head, she stood motionless. I carried our dinner in, placed it down and had only just begun to congratulate myself on my ability to master her stubborn resistance when, in a single swift motion, the pizza was snatched off the plate, splattering its hot savoury topping across the carpet as it landed. Having planted her hooves firmly in the middle of it, presumably to block its escape, she tore a large lump out of it. My attempts to retrieve my dinner resulted in a trail of tomato, cheese and onions being trampled through the house as the pizza, pierced by her hoof, flapped around her ankle.

It was at this point that my ultimatum was delivered. 'I'm sorry, Breeze, but this house isn't big enough for the two of us. Either you go, or I do!'

Bernard was awake the following morning before the alarm. 'I can't sleep any longer, the suspense is killing me,' he stated with a broad smile. As his words echoed up the stairs, 'How is her ladyship this morning?', I prayed for my fairy godmother.

The next couple of days saw Breeze's halo visibly slipping, her height increase and with it her appetite. Anything and everything edible was tried and tested, and all went down and came out the rear without any difficulty whatsoever.

'She must have a digestion like a pig,' Bernard had declared upon finding the fruit bowl devoid of everything but a few pieces of orange rind and a banana skin.

Whilst I admit to having a sweet tooth, mine paled into insignificance compared to Breeze's. She discovered the delights of chocolate. One evening, having polished off a whole bar of fruit and nut I had foolishly left within her reach, she now left no stone unturned in her pursuance of the sweet stuff.

But the day she tried to retrieve a chocolate biscuit wrapper I had disposed of in the kitchen swing-bin, she got more than she had bargained for. I knew without doubt I certainly had, as chaos seemed to be built into

Growing Pains

our daily lives. Breeze's head went into the bin, the lid swung back and as it returned, jammed her head. With a desperate shriek she yanked herself free, pulling over the bin, then like a mad March hare, was out of the door before I had time to issue another warning.

I sighed, drew in a long breath, then muttering to myself, 'Patience, have patience, Marie,' began picking up the soggy, smelly rubbish.

My exasperated yells and menacing looks continued throughout the July days as Breeze ransacked the house.

Bernard, who always manages to see the humorous side of life, was now afflicted with chronic cramp from laughing so much. I guess if it were not for the fact that I now spent most of my day cleaning up after her, I would have found her 'experimental period' funny as well.

This pint-sized, destructive doe dragged cardboard boxes twice her size from room to room, pulled lumps out of the plastic hand-brush and shovel, carried the polythene bucket around the house and shredded newspapers more effectively than any commercial machine could ever do.

It was the last day of July, the day of Bernard's birthday, when Breeze finally fell from grace. I had splashed out on a large box of Bernard's all-time favourite chocolates, Black Magic, and my gift was mouth-wateringly tucked into, then secreted away overnight in a cupboard. The following morning, like Old Mother Hubbard, the cupboard was bare. The chocolates had completely vanished as if by the magic their name implied. Spirited away by a fallen angel, more like! The remains of a chewed chocolate box said it all.

July had come to an end and with it my patience. I can't stand much more of this, I confessed to myself as I scrubbed away at the carpet, which now looked decidedly the worse for wear. My dusting and polishing were next on the agenda, but feeling somewhat depleted of energy, I rose from my knees, straightened my back, and popped the kettle on. I reached for my honey and flopped into the chair for a five-minute break.

Having gulped down my hot sweet tea, I resumed my chores. 'That's the dusting finished,' I congratulated myself, 'only the brasses to polish now.' I paused to pass a hand over my weary eyes, then picked up the decorative oil lamp.

'Peep . . .' Breeze's plea for lunch from somewhere in the garden, coming ever nearer, brought a prickling sensation to the back of my neck. I wondered if we all harboured secret wishes. I rubbed hard on the lamp and waited for the genie to materialise.

On August 2nd, my mother arrived. Due to unforeseen circumstances, namely Breeze, her summer holiday was a whole month later than usual. During the time Bernard was in Carlisle meeting her off the coach, I was filled with a mounting apprehension as to how Breeze would react to a stranger living in the house. I had visions of her fleeing in panic, never to be seen again. We had deliberately deterred visitors, and the humans she had come into contact with had been at arms' length.

A wave of relief flooded through me and my worries melted as Mum alighted from Bernard's car to be greeted with woofs and wagging tails from the girls and a burst of indifference from Breeze.

Mum's visit was to last just two weeks, as a family wedding was imminent, one to which we had also been invited. But her help was to prove invaluable as this summer, being the hottest on record for years, had brought an influx of holidaymakers to the area. None of them seemed to want to go away without a 'sweet-smelling gift from the border country'.

Not only were we still battling to keep up normal

production, but we now had the added problem of having to double it. Mum, being a dab hand at stuffing and sewing cushions, was only too pleased to help ease our work-load.

My mother's first impression of Breeze was one of delight. 'Oh, isn't she dainty, she's all legs and head.'

'She is all tummy, more like,' I wanted to say.

'Oh, isn't she beautiful,' Mum continued, 'I bet butter wouldn't melt in her mouth.'

Not wishing to burst her bubble, I refrained from answering. I had no doubt, by the time she was ready to leave, Breeze would have proved her wrong. I wasn't such an optimist to think Breeze would be on her best behaviour just because Grandma was here. How right I was!

Mum's first evening began quietly enough. Having caught up on family news, she was relaxing completing her daily crossword, until a moth alighted on her newspaper and she shook it off. There wasn't time to warn her that flying objects around the house are best ignored. Breeze had already sprung into action. Mum's mouth fell open in astonishment as Breeze landed beside her on the settee, then proceeded to march up and down it, snatching and leaping at the air. The moth fluttered away, with Breeze giving chase around the room in never-ending circles.

'Well, I've never seen anything like it,' my mother laughed. 'You certainly don't need to go out for entertainment, you have your own amusement here.'

Bernard and I looked at each other and exchanged smiles. 'How right you are, Sally.' There was a mocking note to Bernard's words.

A wind sprang up the following day, scattering our sewing materials every which way. Gathering them up we resumed our crafting in the sitting room, across from the lounge and just off the hallway leading to the garden. This way I could keep an eye on Breeze.

'Something smells awful around here, Marie,' my mother said, sniffing the air disapprovingly.

'Be careful where you walk, Mum, no doubt Breeze has left her calling card somewhere. I'll go and get the brush and shovel.'

Mum's face registered a look of incredulity at the sight of me on my hands and knees cleaning up after Breeze. 'Do you allow her to mess in the house, Marie?'

'Well, there's not a lot I can do about it, Mum, she's a wild animal. I can't house-train her, can I?'

'Mmm, I suppose not,' she replied, shrugging her shoulders, and ploughing back into the pile of fluffy stuffing. This was to be her first insight into our life with Breeze.

The weather, alas, changed over the next few days, with intermittent rain at first, then more persistent. Bernard's time was taken up delivering crafts to the hotels, and journeying down country purchasing materials from the cotton mills. If it were not for my mother sitting sewing, our production would have come to a complete halt, as Breeze demanded more and more of our time.

Wet or not, Breeze had to graze and needed to be accompanied. Whilst the girls were always ready for a walk, standing around in the rain as Breeze munched was hardly their idea of fun. They soon became bored with the whole procedure, and so Breeze and I got drenched alone.

Mum couldn't come to terms with the idea of me standing in an open field, in the pouring rain for hours on end, chaperoning a deer. Her concern grew with each successive soaking. 'You are wet through again, can't you just leave her to graze on her own?'

'No, Mum, she will just run back in, she needs me with her.'

'Whatever possessed you to take a commitment like this on board, Marie?'

'We really didn't have a choice, we were summoned by fate . . .'

Mum evaded this issue. 'Well, you certainly have your hands full, pet,' she said sympathetically.

The unsettled spell of wet and windy weather certainly didn't help matters, as our permanently open door was allowing the elements in. And, apart from a sodden carpet, Mum was feeling cold.

'How long will it be before Breeze is weaned?' Mum asked, staring in amazement at Breeze rapidly draining her bottle.

'About another three months.'

'Goodness, as long as that, how are you going to keep your little business going? Still, she will be gone before the bad weather sets in, and you can . . . close . . . your . . . door.' Mum emphasised each word as she buttoned up her cardigan against a blast of cool wind.

'Oh, Breeze will probably be with us until next May.'

'May!' she repeated, somewhat perplexed. It was then that I realised I had told her nothing of a roe deer's family structure.

I set about explaining . . . 'So you see, Mum, we can't abandon her to suit us, she isn't equipped to fend for herself, we must allow her to follow nature's course.'

Her head was shaking in disbelief.

'You look surprised, Mum.'

'Surprised, I'm staggered! You and Bernard are going to end up with double pneumonia, combing the fields in all weathers and having ever open doors.'

'Don't worry about us, it's Breeze you should be concerned for. The poor little soul doesn't know she is a deer. My heart goes out to her.'

I'd hardly finished speaking when I heard an almighty confusion of cackling and squawking from the yard. As I readied myself to dash out and investigate, Bernard entered, ushering in Emma and Breeze. There was more than a conspiratorial look in the double-act's eyes.

'What was all the commotion, Bernard?'

'Any eggs you get today will be scrambled,' Bernard said with a broad grin. 'Breeze has bullied the hens!'

My smile gave way to laughter. My weird sense of humour was showing itself again. The funny side of me emerged at the oddest moments.

The following day the wind dropped, and the midday sun shone with such intensity it would have melted a glacier. The girls lay under the canopy of the oak tree snoozing, whilst Mum and I merrily chatted away, sewing at the same time.

Bernard's appearance, balancing a tray enquiring, 'Tea on the lawn, madam?' brought forth oohs and aahs of delight as Mum's eyes rested on the plate of cream-filled pastries. I know where my sweet tooth came from, I thought, watching her select the biggest, squelchiest cake on the plate.

Bernard smiled knowingly. 'Enjoy your cake, Sally.' He returned indoors to tackle a mountain of paperwork.

Breeze suddenly awoke, stretched and moved towards Mum's plate. I knew the inevitable was about to happen by the mischievous twinkle in her eyes and her twitching nose. Faster than a wink of an eye, Breeze snatched the cake and I swear she swallowed it whole.

Mum's startled expression as she exclaimed 'Breeze has just pinched my cake!' had me spluttering with laughter. She looked disdainfully at Breeze but soon changed her mood at the sight of Breeze's lip smacking, cream plastered all around her chin. Mum laughed so much, tears were running from her eyes. 'I never even got a sniff at it,' she roared, as we both fell about.

Breeze's antics continued to be the topic of conversation over the next few days, at times pleasing and others vexing. I wasn't exactly enamoured with Breeze's insistence on pulling my hair, for instance.

Every so often, when I least expected it, she would march over, fill her mouth with my locks, and tug. Whether she thought my blonde hair was a bunch of straw, I don't know, but having one's hair torn out by the roots is not exactly pleasant.

Another annoying trait was her plucking of the fine hairs on my arms. Even covering them didn't stop her attempts. She nibbled and pulled at my jersey with her needle-sharp teeth until it looked decidedly motheaten. When she wasn't eating me, she ate Emma!

She had developed a strange habit of licking Emma's eyes, and not gently either. We lost count of the times we pulled her head away from Baba's face, but with her usual insistence on having her way, she returned to annoy Emma the moment our back was turned. Baba's patience finally ran out one evening when, in a final gesture of frustration, she snapped at Breeze, nipping her nose. This was to be the last time she ever licked Emma's eyes.

Mum's holiday finally came to an end. She departed with a suntan and a look that suggested we had taken leave of our senses.

Two-week-old Breeze having a picnic lunch from me in the paddock

Bernard with Emma

A four-day-old Breeze takes her first tentative steps in the rockery under Bernard's protective hand

Five-day-old Breeze finds security with Emma

I hold Breeze at two days old in the palm of my hand.
She weighed less than a bag of sugar

Breeze's first encounter with snow

Breeze shares the indoor comfort of Emma's bed

Emma watches over baby Breeze, concealed in the rushes in the paddock area

I console Emma on the hill in the paddock as she scans the countryside hoping for Breeze's return

Emma on guard duty while Breeze forages in the woods on an autumn morning

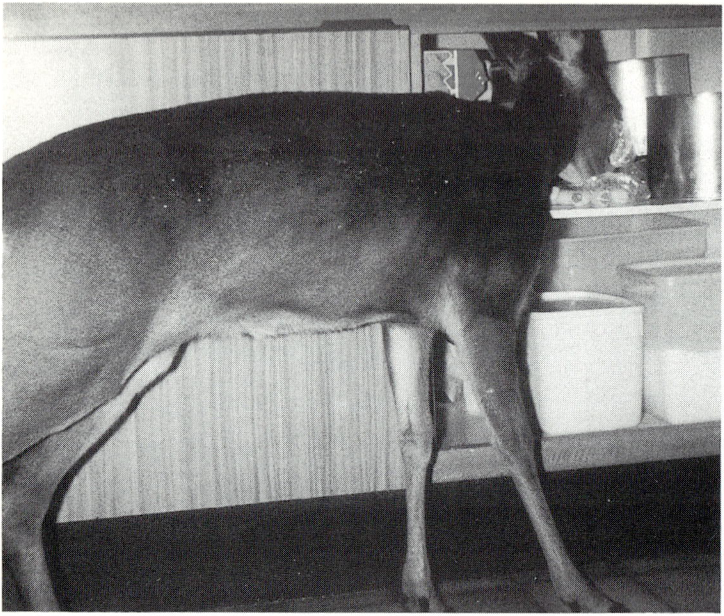

Breeze opens the kitchen cupboard in search of an afternoon snack

Breeze's first Christmas at Coldside: having eaten the
chocolate bells, she contemplates eating the Christmas tree

Emma leaves her supper snack to enquire,
'have you got your head stuck again, Breeze?'

Breeze, eighteen months old, takes a walk, alone,
across the field in front of Coldside

'Mr Twonks' comes a-courting. Breeze's first suitor
advances on the garden on a late July afternoon

Breeze, twenty-one months old, watching the
snow melting in the spring sunshine

13
'Scours'

The following morning, Charlie returned and we were relegated to the front garden once more.

Happy in the knowledge that the girls were not in danger from wandering stock, nor suffering from sunstroke – thanks to Bernard's ingenious canopy suspended by ropes from an upstairs window – I could now catch up on a backlog of cutting out. This was my specific task, as I had a way of juggling the templates to suit the pattern of the material according to the particular craft item. I was actually pleased to be indoors where the air was cool, as the sun tended to sap my energy.

'Aren't we eating today, then?' Bernard's enquiry took me by surprise.

'Goodness, is that the time!' I exclaimed, glancing at the clock. The morning had flown by.

I began preparing lunch whilst Bernard went to check on the girls.

'Marie, come here a minute,' Bernard called, pointing towards the garden. 'She's murdering your fuchsia!'

'Oh, don't worry about it, Bernard, I've given up on the garden, I now just grin and bear it.'

Bernard shrugged his shoulders, then in a loud voice announced, 'You're welcome to what is left, you greedy little piggy, Breeze.'

'Come and have your lunch, Bernard,' I spluttered through my giggles.

The sun had long passed its meridian by the time my scissors transformed the last piece of material into butterfly-shaped cushions. I straightened my hunched back, drew circles in the air with my neck to release the knots, picked up my straw hat and sewing items, and flopped down beside Jester. Her only acknowledgement of my presence was a raised eyebrow and a brief glimpse of 'red-eye'.

Breeze was preoccupied, indulging in the luscious, cascading, cyclamen-coloured flowers of an overgrown fuchsia bush, pausing only to inspect an alighting butterfly with considerable curiosity. Her baby coat had now almost disappeared and was taking on a chestnut hue. The camouflage her dappled colouring had provided in her formative weeks was no longer needed, as she was now fast enough to evade any predators. What a wonderful mother nature is. She not only furnishes her many charges with seasonal attire, but she also equips them with the necessary safeguards to ensure their every chance of survival.

There is no wild creature more beautiful to my eyes than a roe deer, I thought, watching Breeze making short work of a rose by way of a change from the fuchsia. I certainly wouldn't need to prune the bushes this autumn! I looked up into the blue hazy sky, completely cloudless apart from one hovering on the horizon, then switched my gaze back to Breeze. Her little belly, well laden, was now resting and digesting. There was a stillness about the afternoon that was almost tangible, as I watched the lonely cloud centre overhead. Lulled into a false sense of security, I failed to notice it was an ominous cloud, about to shatter my little paradise.

'Scours'

A crisis made itself evident the following morning. It sprang fully fledged from nowhere, and caught us entirely unprepared. I'm one of those people who is totally useless in a crisis. Whilst I have no problem coping with an accepted situation, and am a patient nurse, in the face of adversity I fall to pieces.

'Oh, my God! Bernard, come down quickly,' I screamed up the stairs. 'Stay calm . . .' I told myself. ' . . . don't panic . . .' Then my brain registered malfunction.

I felt desperate as I waited for the vet's surgery to open. The key had hardly finished turning in the lock, when I burst through the door almost knocking Lesley off her feet.

'Hello, Mrs Kelly, how is your little deer coming along? I was . . .' I didn't give her time to finish her sentence.

'It's Breeze I've come to see you about,' I blurted out. 'She is badly scoured . . . can't keep anything in . . . it's running out of her like water.'

As I paused for breath, Lesley enquired, 'How long has she been scoured?'

'Since this morning.' I blinked as my eyes misted.

'Have you still got her on goat's milk?'

'Yes, but it's passing straight through her, poor soul.'

I felt a twinge of sympathy in Lesley's voice as she asked, 'Is she taking her bottle readily?'

'Oh, yes, as eagerly as always.'

Lesley hesitated for a second, then asked, 'Is there any distension of her tummy?'

'No, not that I have noticed.'

'Would her pen have any waterlogged area in it at all?' Lesley further asked.

'Oh, she isn't in a run, she free-ranges.'

Lesley's eyebrows raised, as if surprised. 'Is it possible Breeze could have eaten bog grass or grazed in very wet pastures, if she is running free?'

'Most definitely not, as she is never out of our sight.

We graze with her, well figuratively.'

Lesley's face registered surprise. 'Do you really . . . ? In which case we can probably rule out liver fluke, she would have a pot-bellied appearance anyway.'

'Liver fluke! What's that?' My voice was almost falsetto.

'It's a parasite which infests the liver.'

'Does the parasite live in boggy areas then?'

'Actually, it passes through a small snail which can only live in water. It's a deadly disease, and deer are susceptible to it.'

A cold hand of panic clutched at my heart as I imagined Breeze on the point of death.

'But I'm sure it isn't liver fluke,' reassured Lesley. 'It's more than likely scours.'

My balance was restored once more.

'As you haven't changed her milk, it's got to be some kind of herbage which her digestive system can't tolerate. How about the forest next to you, does it have any rhododendrons?'

'I don't think so, but she hasn't been in the forest, Lesley, we take her over the fields or she is in the garden.'

Garden! The penny dropped. Fuchsia!

'She shouldn't be feeding from the garden,' Lesley exclaimed. 'It's highly likely this is where her scours have stemmed from. The forest is the place for her really, more of a natural diet.'

I refrained from replying, not wanting to go into the great debate of time.

'Now, we haven't anything specific for scours in deer, so I have to give you what we prescribe for sheep in a diluted form. Do you know Breeze's weight?'

'I haven't a clue, Lesley, I only know she is fourteen inches in height.'

'Goodness, she is tiny. How old is she?'

Eighty-nine days, I almost said, but in order not to

'Scours'

sound obsessional, replied, 'Just coming up three months.'

'Oh, as young as that, I'm not sure . . .' Lesley broke off. 'I'd better confer with a partner as to the strength of the mixture.'

Lesley returned with a brown bottle labelled 'For Deer', a syringe and a large sealed packet. 'Ken says she will have to be kept off milk for forty-eight hours. Substitute her bottle feeds with this,' and Lesley handed me the packet. 'It contains powdered nutrients. Dissolve half of it in two pints of water for today, and use the rest tomorrow. Make sure she consumes the whole amount, otherwise she will dehydrate. And you fill the syringe with this mixture' – Lesley handed me the brown bottle – 'and drop it down her throat four times a day. Hold her mouth closed until she swallows. She will start to dry up after two days, but continue with the medicine until it's finished. If her breathing becomes distressed or she develops convulsions, ring immediately.'

I swallowed hard before asking, 'Is this likely then?' then braced myself for Lesley's reply.

'No, not if it is scours. It's rather difficult to give an accurate diagnosis. We are assuming her condition is due to something she has ingested, not in her previous diet, and we don't know if this was of a poisonous nature.'

I felt every muscle in my body tense. Her last words were too horrifying to contemplate. Clutching Breeze's medical supplies, I hastened out, with Lesley's emphatic words of advice, 'Keep her out of the garden,' resounding in my ears.

'Hello, Marie, bet it's lovely up at Coldside today,' a chirpy familiar voice commented as I rushed to my car. I wasn't in the mood, nor had I time for pleasantries. I think I managed a weak smile and may have replied 'Yes', and as far as driving home was concerned, I don't remember doing so. My mind was too busy with my imaginings of Breeze's imminent fate. I entered the house, fell into Bernard's waiting arms and as Lesley's words came tumbling out, I collapsed in a flood of tears.

My tears and Lesley's diagnosis over, Bernard responded immediately. With a look of icy contempt, he strode across to the fuchsia, brandished his saw and without any compunction, lopped off the branches, gathered them into a pile and set the whole lot alight.

If the destruction of the fuchsia had been a cut and dried affair, the administration of Breeze's medicine was not. That was another matter entirely.

Whilst I hardly expected her to part her lips in readiness, I had hoped she would at least remain still and allow me to coax the syringe into her mouth. Instead, she tossed her head like a wild stallion, with an indomitable persistence. Breeze was skilful at getting her own way – she only had to look at me and I would give in – but this was one time I had to stay resolute.

'Come on, Breeze,' I said encouragingly, 'it will make you better.'

She took no heed of my advice. Each time I advanced, she backed away. I had long since learned that it was no good rushing things with Breeze, but as the minutes ticked by I began to weary, particularly as the medicine was beginning to solidify in the syringe. I made one last brave attempt which proved to be useless. It was no good, my patient was most unwilling. We had to employ force. Bernard knelt in front of Breeze and lovingly stroked her face and ears, then held her shoulders firmly. She braced every muscle in her body and looked up at him with pleading liquid eyes. Bernard has always found the act of unkindness unforgiveable, and whilst he knew that what we were doing was in Breeze's best interest, he was fully aware that Breeze did not. How could she? Animals cannot reason.

Bernard's reluctance to cause Breeze distress was

obvious as he managed to part her jaws, but her teeth remained firmly clenched.

'Push the syringe through her side teeth, quickly, Marie.'

Bernard sensed my timidity as I tentatively raised the syringe to her mouth.

'Don't hesitate, just do it.'

Bernard's penitent voice matched his expression as Breeze kicked out her hind legs in protest. Moved by an impulse to heal the hurt showing in his face, I clenched my teeth and depressed the plunger. Bernard tilted her little head backwards, whilst holding her muzzle firmly closed. I, of course, was doing my usual impersonation of a jelly that hadn't quite set, as the trembles in my hands spread around my body. Only when Breeze's throat muscles contracted into a swallow did Bernard relax his hold.

Breeze expressed her immediate disapproval with a squeak and a grimace before springing into the air like a jack-in-the-box, racing around and around in circles, her tongue darting in and out.

'It can't be that bad, Breeze,' I declared, touching the tip of my tongue with a trace of residual medicine. It could. It was! Certainly not flavour of the month. It was downright repulsive. No wonder Breeze was distraught, she was trying to rid herself of the loathsome liquid coating her tongue.

'Poor little soul, I'll make her a bottle to take away the bitter taste,' I said.

'I wouldn't say no to a cuppa to take away mine.'

Bernard's coffee request came out hot and sweet whilst Breeze's bottle of clear lukewarm liquid remained X the unknown.

'I only hope this milk substitute is more palatable than the medicine,' I murmured.

Breeze was halfway through the bottle before a look of bewilderment spread over her face and she withdrew.

I wiggled the teat against her lips, she sucked for a moment, stopped and looked up at me, her eyes pleading for understanding.

'Lovely dinner for my Breeze, good girl, drink it all up,' I implored, gently trying to coax the teat into her mouth again. Breeze remained tight-lipped, then squeaking her resentment, sought sanctuary behind Emma.

Bernard was the first to voice our thoughts, and there was an imperious note to his words. 'I'm afraid we are going to have to be tough and not let our heart rule our head, otherwise she will dehydrate and weaken.'

The obvious was far too terrifying to contemplate. What I felt was too deep for speech. Bernard reached out his hand and taking mine, held it tightly as my eyes blurred with tears.

'I know, pet, it just isn't fair, she was coming along great, now look at the poor little girl.' Bernard's voice, full of pity, had a croaky throaty sound to it.

'Caring makes one so vulnerable, doesn't it, Bernard?'

He nodded his head, avoiding my eyes and in a voice almost a whisper, said, 'You can't legislate for the heart.'

A few days later, Bernard replaced the telephone receiver, turned to me and said, 'Mitch says it's okay.'

Bernard's request to remove a section of fencing separating his field from the forest and to replace it with a small gate had met with Mitch's approval. As the onus is on the farmer to fence his stock off from Forestry Commission land, Bernard assured him he would make it stock-proof. As far as the Forestry Commission were concerned, we kept mum. We would now have immediate access from the paddock at the back of the house, without having to walk to the hurdle at the end of the lane.

The time had now come to introduce Breeze to the forest.

As Bernard's cutters snipped away the wire attached

'Scours'

to a fence post, the girls' excitement mounted. He stapled the raw edges around a length of timber and secured it to the post top and bottom with sliding bolts. We only had to pull back the bolts to gain entry, and close them firmly behind us.

Jester led the expedition into uncharted territory, ecstatically sniffing the unfamiliar scents. Jester entered a world of her own when new smells dictated, her tail swishing in intoxication with each new discovery. Sheba and Emma followed, their noses weaving a path for Breeze through the long grass, which was twice the height of her. We humans didn't fare so well, and the uneven ground was making walking treacherous. We fought our way through the tangle of bracken and stumbled over tree roots in an effort to keep up with the rest of the pack.

Finally, the dense conifers gave way to an area of graceful larches growing out of several layers of pine needles, making a soft springy carpet underfoot. The scent of pine surrounded us. Emma paused to sniff at each clump of grass that Breeze gave particular attention to.

'Oh, the girls are loving this, Bernard.' I was smiling with pleasure, but I became distressed again as a jet of excreta gushed out of Breeze. It's all my fault, I thought miserably. If I hadn't been so concerned with fulfilling our craft orders we could have spent our days in here weeks ago, and Breeze wouldn't now be ailing.

'Plop!' A frog landed on Breeze's back. Her instantaneous bound at breakneck speed caused her to run headlong into a tree, where she collapsed in a daze. Poor Bernard, his timing was all wrong. As he rushed to her aid, Breeze leapt up. Bernard didn't, and he caught his foot in a tree root and fell flat on his face.

'Ooh . . . yah . . . bloody nettles!'

As his cries rang out, he was immediately attended to by a rather rough-handed rescue team seemingly, to all intents and purposes, hell-bent on finishing him off... adding insult to injury by depositing soil over his face... whilst Jester flung out a paw thumping him in the ribs, being her effort to roll him over.

Bernard spent the evening having repeated facial applications of camomile lotion, whilst I, with a thumping pain in my head, continued to be at the mercy of my emotions, scrubbing repeatedly at the green-stained carpet as Breeze continued to splash-down.

I slept dreadfully, the events of the day crowding my mind. With the dawn's pale rays and an imperial headache, I crept downstairs. Breeze leapt up eager for her breakfast, 'peeping' and discharging a stream of green liquid at the same time.

The day progressed with what seemed an endless administration of medicine. Each time I approached Breeze, syringe in hand, she would shy away. She found no solace in her bottle, as even this liquid was now being propelled down her throat. Little wonder my headache hung about all day.

With the sun almost gone and the birds roosting in the trees, we gathered Breeze's grass for her evening meal, leaving the forest to the creatures of the night. My housework and our craft-making had remained undone, and our tummies were protesting from the lack of food. The day had left little time for anything other than Breeze.

Twilight closed in as I filled the syringe yet again. My turmoil manifested itself in impatience. 'This medicine is useless,' I bemoaned peevishly. 'It hasn't helped one iota.'

'Be patient, Marie, give it time. The vet said it would take two days, didn't she?'

'But it's *been* two days, Bernard.'

'No, love, it's been thirty-six hours. I reckon by morning her runs will have stopped and you will be back on your milk round.'

Bernard's casual air did not deceive me. By midnight I had worn my trembling nerves to shreds as my concern deepened and grew for Breeze's well-being.

That night was another sleepless one, and when I did sleep, I had nightmares. I was glad when it was over, but afraid of what the morning would bring.

Pouring disinfectant on the bespattered carpet, I flung open the door and complained, 'This place is beginning to smell like a byre! I don't know how much more of it I can take.'

Bernard, looking decidedly wounded, came instantly to Breeze's defence. 'She can't help it, the poor little soul has no control over her bowels.'

'I'm so sorry, Bernard, I really didn't mean what I said, it's just that I'm so very frustrated. I've done everything I can and it isn't enough.' Down came my tears. Bernard pulled me to him and held me tightly in silent sympathy.

By noon my tolerance had reached its limits. I rang the vet.

Lesley's words knocked me for six. 'I'm afraid we don't know much about roe deer. There's really nothing more I can do other than suggest you double the dosage of scour mixture. It's a case of trial and error really. If her condition remains unchanged by this time tomorrow, however, it's possible she has some kind of infection. See how she comes along with the extra dosage, and we'll take it from there.'

With tears streaming down my cheeks, I ran upstairs and flung myself on the bed. The sound of my sobbing brought Bernard up the stairs three at a time. He sat beside me on the bed. 'Now then, what's all this about?'

My words came out in short gasps. 'Breeze is going to die!'

Bernard sighed and shook his head. 'Let's get one thing straight. Breeze is not going to die!'

'She is, Bernard, she will waste away like that elephant.'

'Which elephant?' Bernard's enquiry had more than a hint of frustration in it.

'That one in the television documentary about the lady who tried to rear an abandoned baby elephant, and when she was six months old became scoured and nothing would stop its scours, and it finally died.' My sobs mounted. 'Why did this have to happen to Breeze, why, why?'

Bernard pressed his finger lightly over my lips. 'Now listen to me, it doesn't matter why... The most important fact is that you refuse to accept Breeze could possibly suffer the same fate as Isha, the elephant. She loves you, and trusts you, and more importantly she wants to live. Now don't let her down.'

I buried my face in the pillow in an effort to smother my sobs, which only served to distress my breathing more.

Bernard's voice dropped a couple of octaves. 'Come on, pet, don't upset yourself like this, it won't do your health any good at all. And quite honestly, just watching you is taking years off my expectation of life.' Bernard pulled me gently to his chest and as we embraced, our tummies rumbled in a combined protest at their state of emptiness.

'I know it's afternoon, Marie, but shall we have breakfast?'

My concern for Breeze had overwhelmed all other feelings. I hadn't given a thought to food. Bernard's sense of humour filtered through my tears, and I managed a weak smile.

'That's better, my love, come on, let's have no more tears.' He brushed my damp hair away from my eyes and affectionately stroked my head, and my state of agitation gave way to one of calm. Suddenly, I had an appetite.

'Scours'

My equilibrium depended to a great extent on Bernard during the remainder of the day. As Breeze's recycling process continued, I administered her medicine with fierce determination. As the last rays of sunshine outlined the horizon with a crimson glow, I thought, 'Red sky at night ... shepherds' delight'. Wasn't that the saying? My mind had no room for doubts, it was too full of optimism. All the signs pointed to a perfect tomorrow.

How often reality disappoints, though.

I had expected a different scenario in the morning, but nothing had changed. I took a deep breath and counted silently to ten, but the anguish still came flooding back. It was a morning of clasped and unclasped hands, of anxious waiting and hoping that Bernard would return from the forest with good news. All the while, that gnawing feeling was eroding what little confidence I could muster.

It was almost noon when Bernard returned from the forest, downcast. The plunger full of medicine was pressed through the side of Breeze's mouth for the third time that morning and by then I was virtually convinced I was only delaying the end. Try as I might, I could not rid myself of the conviction that Breeze was on her last legs. There was nothing more I could do for her. As Bernard took off once again with the girls and Breeze into the forest, I conceded I was defeated. It was time to ring the vet.

As the number rang, agitation crept in. 'Come on, come on, why isn't anyone answering?' I seemed to wait an eternity, then glancing at the kitchen clock I realised why. The surgery would be closed for lunch!

I had another agonising fifty minutes to wait.

Pangs of conscience accompanied me as I trudged into the forest, to Bernard and the girls. 'If only I hadn't let Breeze run riot in the garden ... if only ...' I swallowed hard and took a deep breath, blinking threatening tears away. It wasn't any good crying over spilt milk, the damage was done now.

'Did you get the vet? Is she coming out?'

I shook my head. 'It's their lunch hour.'

'I know you're not in the mood to discuss business, Marie, but I think a few apologetic phone calls to the shops and hotels expecting consignments, would be in order, don't you?'

I remained silent, fearing my voice would betray my total lack of enthusiasm. These last three days my craft room had remained unoccupied. Every waking moment had been devoted to Breeze, and all to no avail.

'Shall I tell them that their orders will be with them by the weekend?' Bernard asked.

As I nodded in agreement, Bernard departed to make the calls.

Wandering aimlessly with the girls through the trees, my eyes glued to Breeze, the time ticked unhurriedly by, seeming like hours instead of minutes. All the while my head throbbed with a pain so acute it had me feeling strangely outside of myself. I massaged my temples to ease the tension, which was mounting by the second as Breeze emptied her bowels yet again.

Still hanging on to a thread of hope, I walked over and inspected it. Convinced my eyes were playing tricks on me, I checked again. I felt a glorious wave of joy. Breeze had passed what could be euphemistically described as a soft sausage.

I tore through the woods, flew through the yard and fell into the kitchen. 'Bernard, oh, Bernard,' I gasped, half laughing, half crying.

Startled, Bernard had dropped the phone. 'What on earth's the matter, Marie?'

'It's Breeze . . . her motions are solidifying . . . she has just . . .' I didn't get time to finish. Bernard had vanished through the door and was making for the forest at full speed.

'Scours'

This slight change in Breeze's motions, which had kindled the smallest of hopes, proved to be the turning point. During the remainder of that day and the two days that followed, any doubts I had harboured as to Breeze's demise evaporated with every cheer-raising defecation.

Breeze's days of being chastised for disgracing herself were gone, forever.

14
Helping Hands

The letter that went off to my sister the following day read, 'Regretfully we shall not be able to attend Tracy's wedding. Our social life has been cancelled due to being irretrievably committed to Breeze.'

Breeze was all time-consuming. Our days left little space for anything else. Day in, day out, we walked the forest, encouraging Breeze to eat a deer's natural diet. Night after night, we frantically crafted away. By the end of the week, with our energy fast diminishing, we summoned a second wind in an effort to catch up with back orders and make a few pennies. With our reserves so depleted, it was evident that our current output was totally inadequate. As we struggled to keep abreast of the current situation, we now had the added problem of Christmas orders coming in.

We could no longer behave like ostriches, we had to find a solution. It was September 8th, the following Sunday, when it came to me. As the girls dozed in the shadow of the bracken and pine, my mind systematically worked out a business plan. As Bernard lumbered up the stairs with what seemed like a ton of cushion stuffing in readiness for the night's work ahead, I put forward my proposal.

'I've come up with the answer to our production problem. We could take on out-workers . . . What do you think?'

Bernard didn't need to consider my question. His favourable answer was immediate and not the least unexpected on my part. Bernard had always valued my judgement in business ideas.

Figures and percentages swamped our minds over the next couple of days. Inevitably the out-workers' wages bill would substantially reduce profit margins, and when final analysis hinted at a fifty per cent loss of revenue we momentarily wavered. It was a 'peep' from Breeze that put us in otherwise determined mood. As it stood at the moment, we were battling to bring in half the usual income, and with Breeze's dependence on us growing by the day, we could visualise an even worsening situation of our finances.

It took team work to complete the final business plan, and as Bernard read it back to me, I smiled in satisfaction. Whilst prices would remain as they were to our regular customers for the remainder of this year, any new accounts would carry an increase of ten per cent. As the crux of our business venture lay in increased turnover, extra outlets would be essential. This was to be my department, as it was an aspect of business that I thoroughly enjoyed. In addition to approaching and securing five extra hotels, I decided to sell direct to the public by way of craft fairs. The latter, being a nett return, would increase our profit margin substantially.

Fired with enthusiasm, and never one to let the grass grow under my feet, the following morning I grabbed the car keys and set off to place the advertisements for 'home sewing staff' on the local shops' notice boards. Things moved very quickly from then on. The telephone rang incessantly. Never had I envisaged such response.

Helping Hands

With replies coming in fast and furious my days were now spent journeying to and fro vetting candidates, whilst Bernard continued to make inroads into the forest.

My years of business experience had taught me that, in the natural order of things, only one in every three applicants came up to expectations. Bearing this in mind I drew up my shortlist. The following week, all twelve candidates were given individual personal tuition in their chosen craft item and, having been allocated materials with which to complete their quota, began their trial period. As the rate of pay was per item, therefore they were paid according to results, I stressed the importance of quality over quantity. In fact I *over*-emphasised, driving my message home that regardless of how keen their interest might be, if they hadn't the appropriate skills, they could not be offered employment.

Mid-way through the following week, as arranged, I called to assess their workmanship. My first point of call was to one of the cushion-fillers.

'How are you, Pat, any problems?'

'None at all,' she replied confidently, 'it was a doddle. In fact I stuffed them in half the time you stated it would take me. Come through,' Pat beckoned, 'I've laid them all out on the dining-room table for your inspection. What do you think?'

My initial reaction of astonishment at her having completed so many was followed by alarm. Results were one thing, but the standard was another. Not only were they understuffed, they were lumpy, and the corners hadn't entered into the scheme of things at all. If this was to be representative of all the out-workers, our plan would never come to fruition.

It was three in the afternoon when I finished my last assessment, by which time I had made a firm commitment to five of the original twelve ladies. Our work force comprised one cutter, two machinists and two hand-sewers. Alas, the second of the anticipated cushion-fillers did not come up to scratch either. Her attempts had the cushions so over-stuffed they weighed a ton and could well have qualified as pouffes.

The last thing I did before setting off for home was to place vacancy cards for a cushion-filler with the respective shops once more. This was a hitch I could well do without, knowing that until a suitable person was forthcoming, I would remain chief cushion-stuffer.

To ensure the continuous smooth running of our business venture, we had decided on defined roles which each of us would perform, and this was not accounted for on my job specification. As Bernard's strength lay in organisation and communication skills, it was only natural therefore that he would take care of our little band of helpers, coordinate the running order, and ensure the work flow was continuous. With my fulfilment coming from the marketing and merchandising aspect, it would still remain my responsibility to keep up our standard of workmanship, something we prided ourselves on. This would involve my hand-finishing every article before Bernard packaged them.

Remaining steadfast to our plan, by the following Monday our expansion programme was taking shape.

The task of filling cushions remained a constant thorn in my side. I found myself hiring and firing would-be stuffers with frustrating regularity. Just as I was beginning to think I'd never find anyone suitable, the ideal candidate materialised. Marvelling at the perfection of each and every filled cushion, I conceded I had indeed gained a rival. Without hesitation I employed our most senior helper, eighty years young!

The wheels of our industry were merrily turning as we left September behind. The countryside was mantled in mist when Bernard, having first collected Breeze's 'pinta', did an about turn and began his journey south to a well-known curtain and bedding manufacturers in Manchester. They were having a one-day gigantic end-of-range material sale, at bargain prices, and with so many hands now to keep busy, it was an opportunity not to be missed.

The sun was striving to break through the hazy blur as the girls, Breeze and I crossed the paddock towards the woodland for our walk. Strange how mist blankets sound as well as light, I thought, opening the gate and ushering the girls through. When two hundred and thirty-five kilos of canine thudded on the forest floor, the canopy suddenly shook, as the roosting wood pigeons rose with a furious flapping of wings into the swirling mist, then vanished. As three excited sterns beat out a rhythm on the grass, the rudely awakened woodland creatures scuttled through the undergrowth leading twitching noses, glued to the ground, on a merry dance, their hiding places concealed by the poor visibility.

Keeping to the edge of the wood, I walked at a snail's pace so as not to become entangled in scrub and brambles. It was a path I knew well and would take me to a larch plantation. I could just see the outline of the girls and Breeze zig-zagging through the bracken. How wonderful to be blessed with the night vision animals have, I thought, as by now they were just a faint blur in the mist. Then, as a single bark rang out, followed by another, my heart quickened. There was no mistaking Jester's bass warning call reverberating through the forest, conveying no sense of direction as it rebounded off the trees.

'Jester, Jester, this way,' I beseeched, breaking into a fit of coughing as the mist caught in my throat. Straining my eyes in the gloom, my heart quickened still more as Emma leapt out, nostrils dilated, hackles up and lip curled menacingly, in full preparation to attack the enemy. She gave a cursory glance in my direction as if to check I was safe and sound, then, roaring like a lion in answer to Sheba's strident yelps summoning her to the pack, she charged off, with Breeze fleeing at her heels.

Talk about the Hounds of the Baskervilles! I suddenly tensed as sounds of muffled human voices mingled with barking challenges. There was someone in the woods... Newspaper headlines flashed in front of my eyes. 'Forestry Commission employees savaged to death by pack of hounds.' If ever I had shortcomings it was my imagination. It was too vivid for my own good at the best of times and at this point it was running riot. Throwing discretion to the wind, I blundered forward in a blind panic, yelling as I stumbled, 'Jester, Sheba, Emma, this way,' knowing full well that all measure of authority in my voice was masked by my nervousness.

'Oo...h!' I skidded on a mossy hump, went sideways into a tree and fell into a thorny tangle. Irritation was now setting in as I scrambled ever onwards, clambering over fallen trees.

My calls turned to shouts, then screams. All were in vain as the frenzied barking continued. They are obviously out of earshot, I told myself, wishing I could whistle a command like Mitch. By sheer luck I had reached the larches, and miraculously the mist was thinning, allowing shape to fill my vision. I was altogether unprepared for, but more than relieved, at what I saw.

Outside of the perimeter fence, just barely perceptible, were a line of bobbing heads, being pursued inside, by the wildly agitated leaping barkers. Then the clip-clop of hooves on the lane, accompanied by continuous soothing tones of 'Wooh, steady on lass,' took on a more defined outline as the riders edged nearer.

'That will do now, leave!' My angry demand stopped Jester and Emma in their tracks, but headstrong Sheba, pretending not to hear, continued her bout of hysteria.

'Sheba, do as you are told this minute and come *here*!' My angry tone of voice made her think twice, and brought her to my side, not before delivering the last word as usual. 'Stay!' I commanded as the horses cantered into full view, all the while Emma bristling like a broom, continued to growl under her breath.

Throughout all the ranting and raving, Breeze had remained totally unconcerned, skipping gaily along the fence without a care in the world. Then she stopped abruptly, craned her neck almost like a swan, raised her right foreleg and remained motionless for all of ten seconds. Stamping her foot down, she advanced towards the peering inquisitor, her nose twitching in curiosity.

Blue, a gentle slate-coloured border collie, generally took her constitutional with the riders and always gave our garden a wide berth, respecting the girls' territory. However, on this occasion, having scented Breeze, and wanting to investigate further, she rushed in where angels should have feared to tread. The moment her nose poked through the fence, Emma's maternal instinct automatically dominated her actions. Like a jack-in-the-box that had been packed too tightly, she sprang into action. With ears flat against her head and neck muscles bulging, she fixed her eyes rigidly on Blue and shot forward snapping a threat which resounded around us like a sonic boom.

Blue cowered in submission, retreated backwards and scarpered down the lane ahead of the trekking party's quickening steps.

Breeze's expression was one of bewilderment as Emma nudged her rump, grumped antagonistically, and continued her tongue-lashing whilst steering her in a semi-circle. Totally flummoxed as to her offence, the delinquent juvenile lowered her eyes, turned her face upwards and thrust it at Emma, eagerly awaiting the warm caress of her loving tongue. Having kissed and made up, Emma wended her way towards me, tail swinging, head held proudly high with Breeze tip-toeing decorously along at heel . . .

The sun's rays now filtered through the tree tops, and colour returned once again to the woodland. Melodious bird song filled the air as we made our way down a ride, through an area of sycamore, and stepped into fairyland.

I was suddenly spellbound surrounded by millions of sparkling silver, silken threads, intricately interlaced, draping the dew-drenched golden ferns. Burnished oak leaves drifted silently earthwards, fusing with the mottled greeny yellows of the sycamore, creating a woodland carpet that defied human design. Ripening brambles gleamed like clusters of rubies, and the aroma of sweet pine was everywhere. Multifarious fungi protruded through the leaf litter like some rising underworld city, each 'umbrella' different in shape, size and colour. These delectable delicacies had whetted the appetite of some woodland creature – a tooth fairy perhaps – as one had been bitten into and was oozing nectar from its cavity.

Obviously an edible species I thought, or rather hoped, as I knew relatively little about fungi. It would take an expert versed in their botanical character to differentiate harmless from deadly. I could only recall two, from my nature study at school.

Fly agaric, a large red-spotted fungus with white warts, was extremely poisonous, and was so named because its poison was once extracted and used in the making of fly-papers. They flourish among fir tress, and this woodland was no exception, as here they stood, in all their glory. The common and deadly death cap, was the other I remembered. It resembles an edible mushroom, differing only in its white or greenish gilled underside as opposed to the pinky hue of a mushroom. Even now, some twenty-five years later, I could hear the teacher's words of warning to us amateur mushroom

Helping Hands

pickers: 'The rule of thumb is never to taste any.'

I felt almost euphoric, the scene was positively enchanting. Oh, how I wished I had brought my camera. Was that a movement in the shadows? I held my breath and gave a sidelong glance in expectation, remembering that fairies were only ever seen out of the corner of one's eye.

I don't know how many hours the spiders had taken to weave their exquisitely crafted webs, but it took only seconds for them to dissipate as a trio of goblins bounded to my side and the whole scene suddenly evaporated before my eyes.

Jester sneezed, clearing her nostrils of the strands of clinging web, and with it, the last vestiges of my mystical Utopia.

'Oh, look what you have done, you are such a roughhouse, Jester!'

Cocking her large head to one side and with a quizzical expression, she wondered what she had done wrong. After all, she knew nothing about beholding the wondrous sights of nature . . . her pleasures involved smells.

An abrasive sound, like the grinding of teeth, turned me around, then my stomach turned over. The fungi had proved to be irresistible.

'Oh, *no*, Breeze,' I screeched, flying towards her like a mad witch, flicking my hand in front of her face menacingly. Alarmed, she sprang backwards and stared at me with uncertainty, the incriminating evidence still sticking out of her mouth, like a cigarette end.

I advanced slowly and purposefully, my arm outstretched, my heart thumping. My hand shot out quicker than an adder's tongue, and to Breeze's great disappointment, whipped the fleshy stalk from between her lips. After searching my face for an explanation, she lowered her head and moved timorously towards another circle of forbidden fruit. I plucked the nearest toadstool and planting myself between fawn and fungi, dangled the tempting morsel in front of her eyes. She stepped forward daintily, wrinkled her nose, smacked her lips and poked out her saliva-coated tongue.

The white gills of the plucked toadstool stared me full in the face. Whatever had possessed me to pull up this harbinger of death? Which species had Breeze eaten? It couldn't possibly have been toxic could it? My doubt was short-lived, for I banished it from my mind, and got on with the task in hand. Withdrawing my arm, and lengthening my backward stride by degrees, I managed to outwit her. I swear I saw a gnome grinning at the stupidity of the situation, as I stumbled blindly backwards. Safely out of the area, I secreted the fungus in my anorak pocket, turned tail and jogged for home, with Breeze at my heels.

By way of recompense for my tantalising tactics of the morning, I allowed Breeze not only an early lunch, but the indulgence of two bottles. We had begun her weaning process three days earlier, by watering down her milk. This resulted in her craving for vegetation to fill the gap, a craving which lasted all day and night long. Holding a banana in one hand (my lunch), a bucket in the other, we set off *again*!

'You need some manure in your shoes, lass,' my father would joke, 'that'll make you grow.' I didn't, but my lack of height had never bothered me – well, until Breeze arrived, that is. Shuffling along like a rickshaw boy minus his rickshaw, my little legs endeavoured to keep pace with Breeze during her continual browsing and searching for the sour docking which would sustain her throughout the night. Why she loves this plant so much I can't imagine, as its taste to me is like the most acrid of lemons. It takes practically all day to collect a couple of bucketfuls, as its singular small leaves grow sporadically, and as fast as a patch is located it is a race against time to gather in the harvest. My gluttonous adversary rapidly devoured ninety per cent of them, then had the audacity to stick her head in the bucket, determined to finish off the remaining ten per cent.

The sun intensified, radiating a tremendous heat for the time of year, and the flies were now becoming quite troublesome in the still air of the woodland. I snapped off a branch from a nearby spruce and waved it irritably around my head, hurrying from the surrounding conifers to the more favourable conditions of an open grassy area.

A knobbly old oak, clothed in lichen and long past its fruit-bearing days, with odd shaped boughs that grew downward and buried themselves in the earth below, made a pleasant backdrop and offered just the right amount of shade for the girls' siesta. I kicked off my wellies, pressed my back into a hollow in the tree and sank down beside them.

Breeze having scrutinised, sampled and spat out a strip of lichen, brought a smile to my face. Mosses and lichens are supposed to form part of a deer's natural diet, or so I had read. Breeze, now engaged in her pre-rest ritual, was making my head reel. What a performance! Around and around she turned in ever-increasing circles, flattening the grass beneath her. Then, with alternating forefeet, she lightly scraped away the humus from her chosen spot, as if turning back the covers on a bed, only to circle around again before finally plonking herself down, half on her bedding scrape and half on Emma's back.

Breeze defied analysis. Graceful and sedate as befits her species – most of the time – whilst at others she displayed as much decorum as a baby elephant. My eyes lingered on her as I remembered the morning's events.

To my mind there are few more contented sounds than an animal chewing its cud. Calmly moving jaws in a slow lateral grinding movement I find hypnotic, and Breeze never failed to find me a captive audience throughout her interludes of digestion. I reached for my camera which I had remembered this time around and snapped the family group. One of our pleasures in life is to pore over the many treasured memories of our canine family, and our photograph albums since Breeze's arrival were growing by the day. She was so much a member of our family now, I couldn't imagine life without her, yet I knew one day I would have to make this sacrifice and ignore the ache in my heart. A sadness began to steal over me, but I swallowed hard, and forced my mind to concentrate on the reality of the present.

Up until now it had been a question of diet which had determined her survival, and increasingly the question we asked ourselves was how to prepare Breeze for her future in the wild. There was little doubt her journey through life would be perilous, for she had predators to contend with. Although delightful to others, if Breeze showed no fear of potential danger, her innocence would inevitably lead to her downfall. Not only had she viewed the humans on horseback without the slightest suspicion, but her consideration of a strange dog as friend, not foe, was foolhardy. Whilst we were well aware that Breeze's unnatural existence had undeniably made her reliant on humans to accompany and protect her, it was a situation we could not retrieve.

Experiencing a familiar tingling sensation as the blood vessels in my head constricted, I flexed then released my shoulder muscles to relieve the tension. I felt a stab of annoyance for allowing the beauty of the day to be marred by my sombre train of thought. I looked upward, closed my eyes, and allowed myself the luxury of day dreaming.

The noise of a vehicle travelling the lane, then grinding to a halt outside Coldside, brought me back from my musing. Could it possibly be Bernard? I hadn't a clue about the time. I looked towards the girls, now sitting up on their haunches, ears pricked, digesting the engine's whine, for some indication. They fell back into the grass, dropped their shutters and continued their sun-worshipping. They had answered my question.

As the post-box lid shut and the van pulled away, I closed my eyes again and wondered as to Bernard's geographical whereabouts, then drifted off, bathed in the sun's warmth.

Alas, the clouds descended again, and it began to rain, much to Jester's disgust. It only took a single raindrop to fall on Jester's head to send her packing, and the droplets now spattering were a few too many. She didn't wait for me to lead the way out, just sped off, determined to reach home before the deluge. I knew the feather-brain would be back as the gate was closed. Predictably, and still at full-pelt, she met us half way. Throwing a black look at Breeze, then a bark, both clearly expressed her message: 'Stop feeding your face and get a move on.' We quickened our pace, and as we made our exit, closing the beauty of the day behind us, the sound of a car engine braking at our gate assailed my ears. Before Bernard had turned into the yard, Jester was home.

As the rain sustained its downpour for the umpteenth day in succession, cooling the air considerably, we knew summer had finally ended.

15
Narrow Escape

It dawned bright, but before dusk closed in, I was to experience a day of disillusionment that was to lead to despair.

Bernard was breakfasting Breeze, when the telephone rang. Having dispensed with the pleasantries of our respective families' well-being, Amy's next words were weighted with concern for Breeze's welfare, undoubtedly well-founded. I gritted my teeth and felt my grip tighten around the handset, turning my knuckles white and my stomach over, as she continued . . . 'The hounds will be setting off in an hour. I'll phone you when the hunt's over.'

As I replaced the receiver I felt my temperature start to rise, and not without reason. To my mind, taking the life prematurely of any wild animal in the guise of sport is gross cruelty enough, whilst the hunting to exhaustion and terrorising to death for pleasure, is downright sadistic.

What of the pitiful orphans left behind to suffer the agonising slow death of starvation?

Man's instruments of his barbaric butchery, the hounds, are innocent executioners, for these naturally friendly, amiable animals are trained to turn from Jekylls into Hydes, only to please their masters.

Never had I given a thought to Breeze, fleeing terror-stricken as the hounds hunted their quarry out of its woodland refuge to open ground.

I recalled reading an article that claimed hounds frequently kill creatures other than their intended quarry. Lamentably, a particular case in question was when hounds ran amok through an animal sanctuary, and savagely ripped a young deer apart. Ironically it had only just been nursed back to health, after almost losing its life in a road accident.

How indebted we were to Amy for telephoning us.

Bernard ransacked the drawer for his stop-watch, I grabbed a couple of buckets, and we raced the girls, hell for leather, to the woods. The herbage-gathering hour over, we returned post haste to the sanctuary of indoors and not a moment too soon. The howls of frenzied excitement brought a lump to my throat as the horns initiated the baying hounds.

Hours seemed to pass, with the air screaming a thousand cries, before the telephone finally rang.

'That's probably Amy, Bernard, will you take it, my hands are wet.' The ringing had ceased and Bernard was already locked in conversation before I'd finished calling out.

Bernard was quietly absorbed in thought as he entered the kitchen.

'Is the hunt over then?' I asked with eager impatience.

He stared at me blankly for a moment as if not understanding the question, then as recognition filtered through, answered. 'Oh, sorry, love, the call wasn't from Amy, it was one of the out-workers ... Funny, we were just congratulating ourselves on our goals being achieved, weren't we? Well, someone has now moved the goal-posts.'

'*Now* what's the problem?' I snapped, failing to

appreciate Bernard's sense of humour.

His eyebrows rose in surprise. 'Well, there's no need to bite my head off, I haven't done anything.'

'No, but it's hardly the sort of statement to raise one's morale, I'm feeling low enough as it is,' I retorted, pushing another soiled towel into the washing machine, and laying a dry one on the floor. 'It seems like there's never an end to our problems. As soon as one is solved, another takes its place!'

Little did I know that the real problems had not yet even begun.

The girls looked up expectantly as Bernard put his jacket on, then took the hump as he slipped quietly out of the door without them. I was feeling somewhat disgruntled myself at losing our cutter at this particular point in time, knowing full well this task would now revert to me. Still, I couldn't blame Catherine for taking a full-time job. After all, she had a young child to provide for.

Maybe a replacement would be forthcoming sooner rather than later, I thought, but there again, knowing my luck at the moment, maybe I was kidding myself.

By mid-afternoon with the land still in a state of violation, my mood deepened to one of depression. The sight of poor little Breeze as she paced back and forth like some caged animal was tearing me apart. Incapable of understanding the situation, her confused eyes went from me to the girls and to the door, over and over again. I bit hard on my lip to repress the rising sobs.

Our undertaking of raising Breeze suddenly seemed bleaker than I'd ever envisaged. At this thought, the flood-gates broke, sending rivers of tears down my face. Jester's relentless sighs, heavy with impatience, added to the ache in my heart, and had me feeling like a kettle that had boiled too long. When at long last Amy rang, I almost fainted with relief.

One word, 'Coming?', with an inclination of my head

towards the front lobby was all it took. Jester, with a burst of unrestrained energy, was first to the door, scraping her paw up and down it, and none too gently either.

'Wait, wait . . . Out of the way,' I yelled as Emma and Sheba, racing around my ankles, almost took my legs from under me. I flung the door open and out they all ran, or so I had thought. It was only seconds later when Emma and Sheba, leaping and barking in frustration, sought me out.

'Yes, we are going for walkies, but I need my wellies on first, girls.' Emma was practically licking me to death as I pulled off my shoe and inserted one foot into my wellington. 'Just wait, Emma, please . . .' I begged, looking for my other boot, which was coming off second best in a game of 'toss the wellie'.

'Oh, Sheba, give me that!' My outstretched hand was completely ignored as she shook it playfully, jumped back and invited me to a game of tag. 'No, I'm not playing, give me my wellie!' With a sharp tug, I yanked it from her jaws, pushed my foot firmly into it and marshalled the pair of them out.

Jester was having a general sniff around the front garden and Breeze, whom I had expected to be doing likewise, was nowhere to be seen. 'Where's baby Breeze, Emma? Is she in the paddock? Go and find her then.' Emma dutifully dashed off, shoulder charging Sheba out of her way as she did so. Turning to follow them, I saw a movement in the window of the dining room and as it took shape, I was rooted to the spot.

Breeze, hindquarters on the settee, front legs precariously placed on the windowsill, neck stretched like a taut length of elastic, was endeavouring to reach a hanging basket suspended from the curtain rail about a foot above her head.

'Oh my God, *no*, Breeze, *no* . . .' Even as I rushed in, I knew I was too late. With an exuberant leap she had latched on to the ivy that had been eluding her. As the plant toppled out of the basket she pulled back in a reflex action, lost her balance and was sent reeling. And in that split second, disaster struck.

The wire hoop supporting the trailing ivy and welded to the pot, encircled her head, then dropped. Breeze's terror-stricken screech as the noose closed around her throat paralysed me with horror. As the millstone around her neck crashed against the floor, she bolted in a frenzy out of the house, vaulted the garden hedge and vanished from view.

I was too taken aback to act swiftly. My mind was trying hard to convince me that this was not happening, it was only a nightmare, but as the scream building up in my throat finally broke free, reality registered. I clambered over the gate and tore after her.

Breeze, fleeing in a blind panic toward the meadow, the ivy flying in her face as it caught the wind, didn't see the fence until it was too late. Her pitiful cries piercing the air were like arrows slicing into my heart and when finally my weakened legs reached her, she lay quite still, quivering in fear, her tongue lolling out, desperately battling for breath. As the racing beat of her terrified heart filled my ears, I begged myself to stay calm.

Agonisingly slowly, my arms reached out and grasped the noose, and in that same moment, the wind increased in velocity, whipping the ivy into her eyes. Breeze gave a shrill scream and struggled to her feet. Aware of nothing but fear as the hostile plant snarled at her, Breeze stumbled to her feet, kicked out at the ivy with her right foreleg and thrust it in the snare. As I stood shivering with disbelief at what had happened, Breeze began her panic run.

I didn't hear the car approaching. The sound was blotted out by Breeze's beseeching cries as she veered in one direction, then another, running, running, on three legs. About to hasten forward I was suddenly conscious of shuddering breaths behind me.

'Don't chase after her, Marie, you will only terrify her more, *stand still*.' The shock dulled my reactions and wrenched my head around.

'Oh, Bernard, thank God you are home... it happened so quickly... I couldn't prevent...'

Bernard pressed his finger to my trembling lip and stilled my tongue. 'All right, all right, Marie, now don't go to pieces, just keep calm.'

Forced to stand by as Breeze tumbled over and over, only to struggle up and on again, and with each repeated fall, become weaker and weaker, had me teetering on the edge of a breakdown. Drained of all energy, Breeze failed to regain her footing and, totally exhausted, she sank down and remained still.

Dropping to my knees in front of her twisted frail little body, I began to pull her leg gently out from under the wire noose, assuring her in soothing tones. 'Breeze, Breeze, please stay still, Mammy won't hurt you.' Suddenly, and with renewed strength, she began writhing beneath my trembling hands.

'Don't handle her like a delicate piece of porcelain, Marie, pull it off quickly.'

'She's struggling too hard, Bernard, I'll break her leg.'

'Forget her leg, it's strangling her! Christ Almighty! Get the bloody thing off, before she chokes to death.'

As Bernard's words registered, my subconscious must have taken over, or did some celestial power release Breeze from her deathly trap... for I recollect nothing.

Tears of relief rolled down as I tenderly cupped Breeze's face between my palms and kissed her closed eyes unceasingly until her rapid breathing gradually slowed and became deep and steady.

Bernard slipped his hands beneath her limp, cold body, and raised her cautiously into the cradle of my arms. Clutching her protectively to my breast, I rocked her as one would a baby, all the while murmuring my love for her.

Breeze whimpered softly, flickered her eyelids, then stared up at me with intensity. As her lingering gaze penetrated my eyes, and beyond, I knew without a shadow of a doubt that this little waif of the wild was reflecting a mirror image of my own feelings.

Only moments before, I had prayed for this day to be obliterated from my mind, yet now, overwhelmed by this torrent of love, I desperately wanted to hold on to today, afraid of the uncertainty of tomorrow and of memories I'd not yet lived. Animals travel through life, not asking for a tomorrow, just living today. I knew part of me would always travel with her.

As autumn continued, leaf after leaf, gold turned to burning bronze. Our days continued to be full of constant activity and whilst I always rush around like a mad thing, without pausing for breath, the fact that morning was only three hours away when I finally finished my cutting out and succumbed to sleep, was now playing havoc with my ability to think clearly. We became used to facing every morning through a smoky haze as I burnt the toast beyond resurrection, being fit only for the bin.

On October 25th, British Summer Time ended, throwing the hens into confusion once more. The sight of the sun-beds being stored away, the front door closed and bolted and the lounge fire and wood-burning stove being lit, begun Jester's annual attack of the winter blues.

Strangely enough, after the initial catapult into the air as the crackling sticks burst into flame, Breeze showed no fear of the fire, in fact rather disturbingly to the contrary. Her curiosity as the flames illuminated the room gave rise to investigation, and seeing her head almost disappear up the chimney, the fireguard became a permanent fixture.

BREEZE: WAIF OF THE WILD

From dawn to midday, the girls and I travelled from one 'service station' to another, Breeze tanking up her tummy, me the bucket. Bernard, fully engaged keeping the craft conveyor belt moving, dashed from the machinist to the hand sewer, to the stuffer, finally collecting, then home. Taking time out to nibble a carrot whilst preparing Breeze's liquid lunch, he took over the foraging of the forest floor, leaving me free to restock the hotels' shelves.

In the foyer of the Royal Stewart hotel, I paused, bunched up my sweater self-consciously in an effort to look fatter, then braced myself before entering the dining room. One thing was for certain, you could always rely on Freda to be frank.

'Goodness, Marie, you have lost weight!'

'Mmm, perhaps I have lost a little, Freda, but then I walk a lot these days.'

'Little? You're becoming thinner by the day, girl.' Pushing the box of biscuits in front of me and drowning my coffee with cream, she continued, 'You really ought to do something about it, that deer will be the death of you.'

Her statement, frank and emphatic, had me searching for words. 'Well, Freda, I had thought of contacting Weight Watchers for volunteers, know of anyone interested?'

'Away with you, girl...' Brandishing a serviette, Freda made as if to box my ears, 'You are impossible!'

Admittedly, whilst our own stomachs complained of neglect most days, the pleasure we derived from Breeze's weight gain made our suffering from malnourishment all worth while.

Autumn, long celebrated for its beauty, is sadly short-lived in this wet neck of the woods, and as winter drew perceptibly closer, the forest in which we had enjoyed so many hours of tranquillity was now totally transformed. As the wind moaned constantly, the untamed pines were not the only ones to creak and groan. Great danes, whilst equipped to swallow up the ground over short distances, are not noted for stamina, and coupled with their advancing years, Jester and Emma were beginning to feel their age. The time had arrived for their winter normalcy of two walks a day and the comfort of the fireside.

Wind-slanted rain blew through the pines, showering down the yellowing larch needles, stabbing me like little daggers. The earthy smell of damp humus assailed my nostrils for the umpteenth day in succession as, head bowed, and shoulders hunched, I ventured forth each morning with my little disciple. She, still missing Emma's attentive accompaniment, paused frequently to look over her shoulder.

The cone-laden larches, being deciduous, no longer afforded any shelter from the elements, and as the penetrating rain percolated down my spine and my wellies, which had sprung a leak, slowly filled with water, I sought the umbrella of the canopy layer of evergreens. Rows and rows of sitka spruce, uniform in age and height, like a regiment of soldiers standing to attention, form a goodly part of this conifer plantation. How fortunate I was that Breeze had a mental map of the wood, because with no defined paths, there is nothing to distinguish one part from another.

The deeper our incursion, the darker it became, and whilst the dense cover gave excellent shelter, the sharp, spiky, waxy coated needles were totally unpalatable to Breeze. With no undershrubs, due to the lack of light, and little ground flora apart from fungi, I had no option other than to retrace my steps. Breeze's needs definitely came before my comfort, and so I faced the direct assault of the inhospitable weather once more.

The wetter I became, the more I conceded I must be insane, for no normal person would be standing knee-deep in ditches plucking docking, whilst the rain bucketed down.

As I held my withered hands up to my mouth and tried to warm my chlorophyll-stained fingers, thoughts of yesterday made me smile. A once avid letter writer, my news to friends and family was now confined to telephone conversations, and last evening's enjoyable phone call from a friend in Hertfordshire saw us comparing life-styles.

'Tell me, Marie, with all this country living, do you still paint your lovely long nails?'

'Nails, Jean! What nails? They have long been reduced to stumps!' Gone were the days of my regular shampoo and set and manicure. I hesitated to think what the beauty salon would think of me now. Long locks, once carefully coiffured, were screwed back in a pony tail, and I had permanently verdant finger ends, from repeated docking picking.

Come to think of it, what *would* Jean make of me now? It seemed decades had passed since the days of my business lunches with top personnel officers such as Jean.

Our lives had changed irrevocably and as our new values emerged, we had attained that elusive quality of life that so many seek. Rising each morning was exhilarating. Each day was lived with fulfilment, luxuriating in the love and trust given freely by a creature of the wild.

As the wind gusted, almost knocking me sideways, I scraped the sticky stinging mass from my eyes once again. Love is unconditional and knows no bounds, I told myself, as I emptied the water from my wellingtons, gritted my teeth to stop them chattering, then plunged my hands back into the stream of icy water.

16
'Twiddly Twonks'

Winter's emissary, Jack Frost, arrived suddenly, without warning, catching us all on the hop. The woods tonight seemed alive with the scurrying of hedgehogs, snuffling like little pigs, foraging the forest floor in a last effort to fill their tummies before retiring to their winter quarters and hibernation. Oh, how I wished we humans could roll up into little balls and sleep off winter's bitter chill.

I was perished, even my fingers were sparkling with frost, bitterly protesting at my obsessive preoccupation with docking. Were it not for the fact that I had a clear specific goal, a sense of purpose and an on-going commitment to my little shadow, my legs would be carrying me to the warm comfort of the fireside.

Upon the attainment of Breeze's six-month birthday, we had implemented the next stage of her rearing plan and reversed her feeding routine. It now ran from dusk to dawn. Although roe deer's natural movements are predominantly nocturnal, Breeze knew nothing of this, and being a nervous little creature at the best of times, the darkness seemed to engender in her a complete lack of security. The faintest rustle had her almost leaping into my arms, but then, thinking back five days to my first night of unaccustomed blackness, guided only by the light of my torch, I too had felt scared out of my wits and almost fled back home to the safety of Bernard's embrace. With the war of nerves now won, and my senses acclimatised to the dark, I was fast becoming an expert at seeing the woods through Breeze's nose, ears and eyes.

In addition to an acute sense of hearing, deer possess an extra faculty to 'smell on the wind'. This scent detection of an adversary is their main weapon of defence. Observing Breeze looking suspiciously around her, head facing the wind direction, testing it, only to repeat the process a few yards on, set me to thinking how deer ever managed to fill their tummies, being forever on their guard.

Breeze's superior senses awakened in me an extra awareness, and throughout those nights, I glimpsed woodland inhabitants rarely seen in daylight hours. Conifer plantations offer excellent concealment to wildlife: they lie up during most of the day, awaiting the obscurity of dusk and the beginning of the night shift. Deer are always the first to break cover.

My transient previous trespasses through No-Man's-Land hadn't as yet granted me the pleasure of observing its inhabitants at close quarters. My encounters were always at arm's length, and then only when down wind. Hardly surprising, as whilst Breeze made no more sound than a mouse, my heaviness of foot was unavoidable. Even my exaggerated tip-toe announced, 'Watch out, there's an intruder about!' Nevertheless, the thrill of glinting eyes caught in torchlight, vapour rising on the cold night air from shadowy animals on the move, and the occasional flash of a long sinuous body, dark stockinged forelegs and disappearing red flanks were rewards enough.

With dawn's first light even this heady opulence was

'Twiddly Twonks'

surpassed as the daytime creatures, refreshed from their slumber, rejoiced at the re-birth of another morning. To feel the pulse of life throbbing through the woodland was like a regenerative current going through me at the end of my fourteen-hour shift.

The advent of shorter days and longer nights saw the twilight to sunrise shift getting the lion's share of the available food source, but not so tonight. Pausing to change the extinguished batteries in my torch, with fumbling fingers that were fast losing sensation, and forever blowing my nose to avoid the suspended droplets becoming icicles, just about saw my resolve deserting me. The colder it became, the less Breeze ate, and as she recoiled from each and every silver strand of grass, it was blatantly obvious that she disliked frosted vegetation. For all it was only midnight, there seemed little point in carrying on, as locating docking under this heavy blanket of frost was nigh on impossible.

'Time to head home, little one,' meant only one thing to my devotee, now smacking her lips in anticipation of her only remaining bottle feed. Breeze was not as yet fully weaned, but this was next on the agenda.

As the Ice Age continued into its third day, our motoring costs escalated with Bernard clocking up the miles in his search for frost-free areas of vegetation. Motorway lay-bys, grass verges, river banks, nowhere was overlooked in our mission to keep Breeze's rumen operational.

Even the hotels granted my eccentric request with a wink and a guffaw to prune the rose bushes and eradicate the brambles from their gardens. As my hands ran red from thorny spurs biting mercilessly into them, I wondered how deer managed to avoid being torn to pieces. I could only surmise that their sensory glands exuded a natural anaesthetic.

Adjacent to one of the hotels was a public footpath leading down to a river, which I followed faithfully each day, gathering the docking which grew in abundance. The first day a rather bemused lady walking her dog had enquired, 'What are you picking? I'm intrigued.'

'Docking for my deer,' I replied with candour, and instantly regretted as a barrage of questions delayed me for a good half-hour.

The following day, having aroused the curiosity of a fisherman, I replied untruthfully, 'I am a herbalist and use the leaves in my potions,' thinking that would be the end of the conversation. Trust me to have touched on a subject that interested him. I was bombarded with a string of his ailments. When he asked what might best cure his lower back trouble, I could only think to tell him to give up fishing. I felt a bit of a pillock as I hurried away, and made a firm resolve that if ever asked again I would state I was a witch. At least *this* answer would ensure I wasn't detained in my mission.

With our nightly roaming suspended until the weather improved, Breeze's only source of independent grazing was on our morning and afternoon walks. With the grass still frosted, she turned her attention to the seeded heads of rushes slightly thawed by the sun's weak rays, hardly filling, nor nutritious, I shouldn't think. Supplemented with whatever we had gathered, dog biscuits and toast, she still sought more. Even the blue tits' nuts were devoured as she rose on her hind legs and pulled down the containers. Tipping the scales at sixteen kilos she certainly wasn't underfed, for normally this weight would not be attained until eight months of age when, with the advent of spring's new growth of vegetation, the remaining four months to adulthood would increase this to thirty kilos. Our dilemma was that we had a six-month-old deer with an eight-month-old's appetite at a time when suitable vegetation would become even scarcer.

Breeze's initial fright at her own duplication in a frozen puddle turned into a magnetic obsession with

'Twiddly Twonks'

each glassy pool she encountered. Craning her neck, emulating an ostrich, she would stare intently at her image, scrape at the ice, flex her neck in a full circle, stare again, take a standing leap whilst clicking her heels in mid-air, land invariably legless, and skate across the ice on her tummy. This spectacle of stupidity had the girls hugging our sides, somewhat embarrassed to be seen in her company. When Emma looked up at me, her face incredulous at Breeze's folly, I smiled at her encouragingly, 'It's all right, Baba, she's just a twonker.' This gave birth to Breeze's pet name of Twiddly Twonks.

The thaw when it finally came at the end of November, gave way to a chilling blast direct from Siberia, laced with icy rain. The frosts had speedily killed what grass was left, and docking, now conspicuous by its absence, was replaced by watercress bought in by the trayload. Thus began our spiralling account with the greengrocer.

December crashed in with a warning to batten down the hatches, as the howling heartless wind tore through Coldside taking everything moveable in its wake. Dustbins, slates, plant-filled containers, even the poor hens were suspended in space as a wild untamed energy tore the door off the henhouse on its unrelenting journey. The poor little chuckies were scared out of their wits as their raking claws desperately tried to cling on to anything solid.

The rattling, creaking and groaning of other doors, straining at their hinges, were all ignored until the hens' rescue operation was complete. Once safely re-housed in the downstairs bathroom, clinging tenaciously to the shower-rail, we set about securing the other out-houses.

The following hours proved to be a perpetual fight against nature's most lethal weapon, and we felt ourselves fortunate to have escaped with the loss of only one other door.

BREEZE: WAIF OF THE WILD

With the coming of night and the air still terrifyingly turbulent, we now had heavy rain to contend with. It was coming down in torrents as I made the final dash into the paddock with the girls for 'pee-pees'.

From the anonymous black of the night, something rushed into me full force, knocked the stuffing out of me and the torch out of my hand. I wasn't exactly sure what happened after that. All I know is that Breeze bolted, pursued by some flying object. I felt a sudden panic, standing in the pitch dark shouting aloud, 'Breeze, Breeze, this way Breeze,' and nothing but the fierce wet wind, howling back, tirelessly, into my face.

Bernard looked up from his invoices and stared at me in disbelief. 'I've lost Breeze,' I sobbed.

Our search by flashlight of the paddock and field beyond served only to disclose huddles of sheep sheltering in hollows, and a bird box on the move, most likely the mystery object responsible for both my assault and Breeze's flight.

'It's pointless staying out any longer in this storm, Marie. Breeze will eventually find her own way back, and if we leave the back door open she can let herself in.'

I offered Bernard no resistance as by now I was heartily sick of the buffeting wet wind, and was beginning to feel quite sea-sick. The mental and physical strain of the day, coupled with the fact that we hadn't as yet eaten, was beginning to take its toll on me. With a chair wedged firmly against the door to avoid it slamming shut as the gale tore through the house, we built the fire up the chimney and with our hands clamped around steaming hot cups of cocoa, slowly began to thaw ourselves out.

We spent the next two hours sending impatient glances toward the door, with ears permanently cocked listening for Breeze, whilst giving ourselves various reasons for her non-appearance. Slowly we began to suspect the truth.

'Twiddly Twonks'

'I doubt she will come back of her own accord now, pet.' Bernard's disappointed tone as he stared into the dying embers of the fire saw my facade crumble. Thoughts of little Breeze, alone, drenched, frightened and lost, invaded my heart, and I felt an oppression almost too great to bear.

'We can't wait any longer, we must find her,' Bernard said with decision.

Our meals lay heavy on our stomachs, but it was at least fortification for the night ahead. The hostile wind snatched at our faces, and the rain, now hail, stung our eyes. But the weather no longer counted, we were compelled to continue our search.

Halfway down the lane we split up, Bernard taking the meadow area and me the lane and surrounding field. Aided only by a flashlight, my field of vision was limited to a tunnel of light either side, a featureless landscape of various shades of black and the probing circles of Bernard's torch waving erratically. My throat burned from screaming, and a callous insistent voice inside my head repeated, 'You should not have waited so long.' Swallowing back the lump in my sore aching throat, I called out once more, then listened for sounds, but it was useless.

Bernard's distant hoarse cries of 'Breeze . . . Breeze . . .' being whipped up in the wind was all I could hear, apart from my own pounding heartbeat then, for one terrifying second, it stopped as a scampering shadow, hardly discernible, masked by the rain, passed momentarily through the beam of my torch. Was it a running sheep? No, it couldn't be, they were all confined to the back field. I swung the light from side to side, straining my eyes, searching for movement, until my head swam with dizziness. Nothing. Nothing but the wind charging among the trees, like a demon out of hell, and the rain pressing its dark weight down on me. I felt as if the worries of the world had come to rest on my shoulders, putting a damper on my energy.

My leaden legs battled to keep their ground as I slipped and slid in the mud, and my throat burned with each repeated call of diminished hope. My eyes were drawn to the approaching light from Bernard's torch. Our futile search would soon be over. The stark facts stared me in the face as our torches swept the invisible fields.

Suddenly, two tiny pin-points of light flickered in the torch's beam, then vanished. I swung my torch to the right, then left, then right again. The tunnels of light were empty. Had I seen anything? Was I hallucinating? No, I had seen something, albeit not definable, and then there was something else, a feeling in the pit of my stomach that had never proved me wrong. As I argued with myself, my natural sense of optimism began to reassert itself once more.

Suddenly there was a movement in the darkness ahead of me, no longer did my eyes contradict me, for caught in the cross-beams of our torches, was a silhouette, unmistakeably Breeze. A surge of adrenalin elevated me to euphoria, and when I came down to earth, Bernard was approaching with rapid strides.

'It's Breeze!' Bernard cried jubilantly. 'Shine your torch to the left, to your *left*, Marie, not the right.' Typical of me, my sense of direction always left a lot to be desired.

As our beams overlapped, Breeze inched into view. No wonder I couldn't get a pattern to her movement, there was a fence between us and she was running back and forth along it, desperately seeking a way out.

'Keep calling and find the gate, Marie, I'll try and keep my beam on her.'

The gate by which Breeze had presumably entered was a good three hundred metres away, but physical fatigue no longer counted. Whilst my feet flew across the field with renewed strength, I was having difficulty

'Twiddly Twonks'

with my voice. Seized in the grip of a dry throat, it refused to obey orders, but Breeze needed no words, my presence was advertised on the wind, and using her nose to guide her, *she* found *me*.

Her ecstatic 'peeps' as she brushed repeatedly against my thighs turned my heart over within me. Had I not been basking in Breeze's attentions, I would have seen the gate, open, when it entreated my entry, now swinging back on me with a vengeance. The river, which by now had broken its confines, was surging across the meadow, as I lay prostrate on the sodden ground, but I was aware of nothing but Breeze's face, a mask of devotion, staring into mine.

Then, as she lowered herself on her forelegs and assumed a prone position alongside mine, it wasn't the flash flood that was threatening to engulf me, but an overflowing heart.

Bernard's shouting brought me back from my drifting as he approached, waving his torch back and forth across the pair of us. 'Ahoy there, maties!' Breeze sniffed at the air, then raised herself and confirmed her recognition by rubbing up against him affectionately.

'Welcome back, Twiddly Twonks.' Bernard stroked her head lovingly. 'Couldn't you have chosen a better night than this to go AWOL?' Then, casting his eyes over my horizontal form, he declared with impish impulse, 'Good God, Marie, I know that to be at one with nature you must blend with the elements, but don't you think this is taking it a bit far?'

Bernard's refreshingly unpredictable humour turned my smile to laughter, and as it caught in my throat I began a sudden fit of coughing. Clasping his arms around my waist he hoisted me up.

'Just look at us, torn to tatters by the wind, soaked and frozen to the marrow, and probably a case of double pneumonia to boot, and all for the love of a deer.' Bernard squeezed my hand, pursed his lips in a kiss and blew it towards Breeze. 'But you're worth it all, Breeze, you're worth it all . . .'

We awoke to a morning of idyllic still, dressed in white. How the girls loved the snow. They cavorted, snorted and bit into it with the silly delight of puppyhood. Breeze, on the other hand, walking as though on egg-shells, stared in perplexity and with a plaintive note enquired, 'Where has all the grass gone?'

'It's still here, Breeze, you must find it,' I answered, exposing a green patch with my foot. Uncertainly, she rubbed her nose along the ground, then made a scrape with her hoof. What strange feelings must have stirred in her pea of a brain when she first encountered snow.

Before coming to Coldside we could afford the luxury of indifference at the sudden arrival of snow. But when our first winter saw us sealed off from the outside world for five days – the council's priorities being concentrated on major roads and some minor ones but never, seemingly, single track lanes such as ours – we applied the motto, 'Be Prepared'.

Considering we have no road lights, not even 'cats' eyes', with which to manoeuvre the narrow twisting lanes, and that we have to transport our household refuse a quarter of a mile to the nearest collection point, one wonders why rural rates are so high.

The midday weather forecast said one thing, but the snow-laden clouds touching the ground said another. Having run the gauntlet in February, when man and elements were at variance, we now viewed any forecast with scepticism. With hotels to stock up for the festive season, out-workers to visit and animals' tummies to consider, discretion beat the weatherman hands down.

The snow continued falling fast and furious. Knee high, then thigh high and with the trees half submerged, the woods lost their identity. As virgin drifts covered all paths, the girls had to jump to make any progress, and with the wind whipping up the snow behind us, covering

'Twiddly Twonks'

our prints, back tracking became impossible. 'Knee to chinning', our way out of this circuit of confusion, was stamina-sapping, to say the least.

'Use your brain . . . there must be an answer,' I told myself repeatedly. My mind was still searching for a reply as I sank into a dead sleep that night, then somewhere in dreamland I saw streamers. Flake fell upon flake as we tied tell-tale red ribbons around the branches of the conifers. With our way in now ingeniously marked, no longer would we, and precious hours, be lost.

It was on the fourth snowbound day, when Breeze had wolfed down everything green in the house and was refusing to eat the bunches of holly we had laboriously lopped down (a supposed winter sustenance), that we put out an SOS.

Chris and Jean came to Breeze's immediate rescue, kindly stripping the remaining leaves from their rose bushes and gave us *carte blanche* to 'pick your own anything' from their productive, now dormant, vegetable garden. With the addition of Eddie's turnips and sprouts, Breeze managed to stave off her hunger pangs for all of two days. I was well beginning to understand why deer travel for miles in search of food.

It was when Mitch called 'Ninny, Ninny', and his hungry flock ran bleating to the feeding troughs, that I was suddenly struck by another idea. In addition to hay, Mitch feeds his sheep a complementary winter mixture of dried concentrated cereals, fruit, fibre and other nutrients. Why not feed Breeze the same diet! Mitch willingly provided the samples. We won on the swings and lost on the roundabouts, which was no more than expected, as I recalled having read that roe didn't readily accept hay.

'When the weather breaks, Marie, see if Mary has a similar feeding stuff for goats, it may be better suited to her digestion . . .'

BREEZE: WAIF OF THE WILD

By the weekend, a thaw had set in, shrinking the snow and flooding the landscape. As the howling wind, bordering on gales, returned, the ground dried up considerably. According to the weathermen, this was only a temporary respite, forecasting further snow by midweek, and by now the betting odds for a white festive season were odds-on. After three nights of frost and clear starry skies, Christmas Eve saw the snow returning like a rightful claimant to its home ground.

With our craft deadlines accomplished, our larder stocked up, and two sacks of 'goat-mix' awaiting Breeze's ever rapacious mouth, for the first time in weeks we felt blissfully relaxed and we were looking forward to this Christmas Day more than any other.

On Christmas morning we waited excitedly outside Santa's grotto whilst Bernard, behind closed doors, put the finishing touches to the tree. That's what he said he was doing, but his protruding paunch as he attempted to secrete our surprise presents up his jumper gave the game away.

The moment the brass knob turned on the dining-room door it was like sale-time at Harrods, as the girls pushed through as one. Bernard's 'Ho-ho-ho! What has Father Christmas got for his girls?' to the accompaniment of muffled squeaks, as he squeezed the gift-wrapped doggy toys, sent the girls wild.

Jester, with eyes beaming like car headlights and her tail swishing furiously, was chasing Bernard around the room while Emma, gyrating as if she had an invisible hula-hoop around her body, followed suit. Sheba, trying to head them off, barked in frustration, adding to the pandemonium. As Bernard yelled 'Timber!' I jerked my head around. Too late! Breeze brought the Christmas tree crashing to the floor with me beneath it. Crawling clear, pulling silver strands of tinsel from my eyes, I shook my head as if to shake my brains back into position. Feeling my flushed face becoming redder by

'Twiddly Twonks'

the second, I curled my hands into little fists. At this point I might well have been inclined to strangle Breeze.

It was the sight of Bernard contorted with laughter and Breeze having reduced the room to chaos, blithely polishing off the chocolate bells at record speed, that saw my glowering scowl soften and change to a broad grin.

'Oh, Twiddly Twonks,' Bernard spluttered, 'if Walt Disney had seen you first, *Bambi* would certainly have been a different story, isn't that so, Marie?'

I didn't reply, I couldn't, I would have choked on my laughter.

We had viewed February's new growth and mildness of weather with suspicion, knowing only too well the catastrophic consequences should it prove to be a false spring. For just as winter comes early to Coldside, spring normally arrives late, with April showers falling in frozen golf balls or snow.

Lambing began in the second mild week of April. It was then that Jack Frost, the executioner of wildlife, struck. Tadpoles lay in suspended animation in frozen pools, never to be allowed birth. Everything in the garden was nipped in the bud. Hedgehogs lured from hibernation by the prospect of an early spring, found their food source cut off as insects and worms returned to their soily depths, and birds, caught mid-nesting, flocked back to the garden, thankful for any household scraps. New-born lambs, ejected from the warmth of their mothers' wombs into a hostile world of icy white wetness, perished within minutes, never having risen to their feet. Others bleated pitifully, finding themselves lost, abandoned or sadly bereaved, confused and all alone in the blinding blizzard.

As ewes heavy with lamb searched frantically for a sheltering place, their ordeal about to begin, we ploughed through the slush to offer our assistance to Mitch in his uphill struggle. The ewe lying in a water-filled ditch, her bloody tail affirming motherhood, yet not a sign of her offspring, lifted her head half-heartedly. Luckily she still had a vestige of life left. With no time for deliberation when Mitch asked 'Have you got an old piece of carpet, Marie?', I didn't stop to ask why, I went about my mission as fast as my little legs would allow. With the aid of the ingenious harness, Mitch and Bernard hoisted the ewe up and bounced her back to life, before carrying her some distance to Mitch's straw-filled pick-up. As the casualties mounted, with four more flaccid and sodden little sparks of life steaming now on the bottom-most shelves of the slow wood-burning stove, our kitchen had become a recovery room. Those that had already revived were wrapped in woolly jumpers and left cuddling up to hot water bottles. The smell of warm lamb permeated the whole house.

Mitch made a phone call before he departed. 'You now have nine more mouths to feed, Mother!' With the lambing only begun, there was no doubt that nine pet lambs would be but a drop in the ocean.

17
The Yearling

We were not the only ones glad to see the snow finally relinquish its hold. Mitch and his working partner, Tess, were about ready to drop with fatigue. A working collie can lose half its body weight during the lambing season, and Tess was no exception. Breeze, on the other hand, had not lost an ounce.

The renewed burst of spring vegetation came not a moment too soon. Breeze had begun grazing the floral-patterned hall carpet and whilst the flowers would not part company, the fibres did. Talk about eating us out of house and home!

Having first checked my 'things to do today' list, I turned to the forward planner. 'We will need to rearrange our schedule for tomorrow, as I won't be around until midday.'

'Why's that, then?' Bernard enquired.

'Well, it's back to the twilight zone for me now, starting tonight.'

'I think we need to talk.' Bernard's voice was low and gentle, but a hard seriousness lay behind his eyes. He leaned forward and clasped my hands in his. 'Now listen, pet, Breeze's future is dependent on two factors, not one: her ability to be self-sufficient in her food source *and* her independence of us, so no more midnight chaperoning, she must learn to graze on her own in the woods.'

I felt shock, then anguish, and withdrew my hands from Bernard's.

'But you can't chase her away yet, she's not old enough, Bernard.'

'You are jumping the gun, Marie, I'm not suggesting we send her packing, but as roe are solitary and non social normally, she needs periods on her own to develop this side of her character, don't you agree?' Bernard's voice, full of resolute purpose, had me nodding an affirmative whilst my mind tried to sort out all my fears at once . . .

'But she is frightened of the dark, Bernard, and terrified of foxes and would probably run away from her own kind . . .' My words came tumbling out despairingly.

Bernard looked at me in dismay. 'But this is precisely the point I'm trying to make. How are her primitive instincts to awaken when she doesn't even know she is a deer? She probably thinks she's a dog.'

I looked at Breeze, sandwiched between Emma and Sheba. The proof confronted me.

'We have to level with each other, Marie,' Bernard continued. 'Breeze is far too dependent on us to survive the wild, and unless she is made to stand on her own, we are going to end up with a pet deer, and this was never our intention, was it?'

I took a large intake of breath and lowered my eyes. 'No . . . no . . . you're right, Bernard, I'm probably too over-protective. It's just that I was waiting until she could at least leap out of the woods if there were danger, before throwing her to the lions.' My emotions were too near the surface to say any more on the subject. 'I'm going to make a cup of coffee, do *you* want one?'

The Yearling

Bernard wrinkled his brow in a slight frown at the curt tone of my voice before speaking. 'You're not happy, are you?'

'No, not really, Bernard!'

'Well . . .' I could almost hear the activity inside Bernard's head, ' . . . how about we leave her in the woods from noon until dark falls, this will give her a good nine hours alone.'

My mood suddenly brightened at Bernard's placatory words.

'But we must stop picking food for her and bringing it in, don't forget she hasn't many weeks left with us, so she must be taught to be self-reliant.'

Forget! I needed no reminder, for didn't I struggle daily to quell this very thought?

Breeze was tucking into a patch of sweet, green leaves as I stealthily crept away. I felt like a jailer committing her to solitary confinement as I secured the hurdle behind me. I ran quickly across the paddock and dropped from her sight behind the hill, pressing my ears flat to my head to shut out her high-pitched peeps. I slid down the grassy slope, not daring to look back. I spent the best part of the afternoon looking through tear-stained binoculars, hoping to see her grazing, yet knowing from her constant calls of distress that it was a foregone conclusion she was not.

I greeted Bernard's return tearfully. 'This isn't going to work at all. Breeze is far too distressed to eat anything. All she's done for the last three hours is to run up and down the fence looking for a way out. Just listen to her Bernard, asking why we've deserted her. We've got to let her out, or join her.'

'No!' Bernard looked annoyed. 'We've got to stay resolute. We discussed the whys and wherefores this morning. When she is hungry enough she will eat.'

My craft room became my salvation for the remainder of the afternoon, as I sewed in brooding silence and urged the light into darkness. I dropped three fizzies into the glass of water this time, as the previous two hadn't eased the pain one iota. My headache was now a fixture like some kind of permanent penance. I put away the fiddly little hedgehogs, unable to concentrate, and slipped cushions into poly bags instead, as this was a task that needed no thinking about. My mood was still mutinous when Bernard entered the room.

'I've brought you a nice cup of tea, pet, before taking the girls for their walk. I'll go to the meadow so that Breeze doesn't see us.'

I took the cup from Bernard wordlessly.

'See you later then,' he said softly.

I was staring out of the window watching the sun drop lower, when Bernard called from the bottom of the stairs, 'I've brought someone back to see you.'

Into Bernard's words came a familiar 'peep', winging my feet down the flight of stairs. With a sound that was almost a sigh of relief, Breeze began her rubbing ritual on my thigh. She's sprouting by the minute, I thought. It seemed like no time at all since she was caressing my calf, and here we were awaiting only her leaping ability to complete her final stage of physical development.

Within a few weeks Breeze would be termed a yearling, and I could recall every minute of the year with a clarity as if it was yesterday. I felt a twinge of sadness, knowing only too well that the clock was running now. Breeze's tomorrows in the family unit were numbered. I blinked away my tears and clenched my teeth, forcing the thought back into the innermost recesses of my mind. It wasn't only the pressure of Breeze's body that brought a sudden weakness to my legs.

The following dawn of destruction by the hungered insomniac, saw all hopes dissolving of us ever having any heirlooms. I didn't mind the empty coal scuttle, and its lumps of coal licked clean glistening among her droppings, nor did I mind the reek of ammonia, but my collection of hand-painted goose eggs, now just crushed shell, and the beautiful feather-crafted butterflies, plucked of their wings resembling stick insects, I more than objected to.

'Something is going to have to be done about this, Bernard,' I declared in a voice that was less than composed.

'Yes, Marie, what do you suggest?'

The solution had been delegated to me. Having dredged my mind, I knew I had two alternatives, to gather buckets of grasses again or shut her in the utility room. In the end, I did neither.

It was on the third despairing day, with Breeze as far from accepting her own company as on the two previous ones, that I found the solution, and by accident rather than design.

The afternoon was overcast and threatening rain and, not knowing how long it would be before an opportune moment arrived in which to make my quick getaway, I took my anorak. I hung it over a fence post and continued leading her to solitary confinement. I felt like a traitor, introducing Breeze purposely to a flush of irresistible docking, then turning tail and bolting before her inevitable alarm call, when she discovered my deception.

As with the previous days, every hour on the hour saw me creeping up the hill, binoculars at the ready, keeping a watchful eye on the reluctant loner. Today I could hardly believe my eyes as, contrary to my expectations, Breeze was contentedly grazing. With a thrill of deep pleasure, I slipped silently away.

Emma was hovering at the paddock gate, directing an enquiring look my way.

'Oh, it's not time for Breeze to come home yet, Baba,' I whispered somewhat apologetically, watching her tail abruptly still and droop. Emma always made such an immense fuss of Breeze when she returned. Having given particular attention to her bum, and digested the information as to her whereabouts for the day, she would shower her with kisses and, still retaining her rights of protection, mother her for the night.

'Let's go and see Jester and Sheba, shall we?' With a downcast look and a sigh of resignation she followed me into the garden. Emma grumped her request for Sheba to move along and allow her on to the sun-bed. It was pointless asking Jester, on her back in rapt enjoyment, legs in the air, inviting the sun to caress her tummy. The seeming unnaturalness of this position was perfectly natural to Jester, and personified her truthfulness and openness, which at times in the company of acquaintances, never friends, was downright embarrassing.

Bernard's totally uncomplicated 'Big Girl' was showing her age now. Her all but white muzzle and less than spry walk, were to be expected of her ninety-four doggy years. A remarkable achievement, considering great danes are not noted for their longevity, unlike labradors.

I felt a great loneliness well up in me. Joy's life had ended when she was only one year older than Jester. It was a bitter pill to swallow.

Having pegged Breeze's washing on the line, I poured the last dregs of my favourite perfume into the bucket of soapy water, dropped to my knees and began scrubbing. It was a repetitive, laborious chore, but having long accepted that a deer couldn't be house-trained, it was pointless moaning. Holding my head aloft, I tried to distance myself from the sweet sickly aroma. With so many conflicting scents trapped in the carpet, my home was beginning to smell like a house of 'ill repute'.

I felt a surge of conscience as I stared at the clock. It

was almost two hours since I last checked on Breeze. My cuppa would just have to wait. I picked up my binoculars by force of habit rather than expectation, as in all probability she would be deep in the forest. Just as this thought crossed my mind, another followed. What if she didn't come to my call when dusk fell? The bottle which had always been my trump card, I no longer held. I certainly couldn't leave the hurdle open for her exit, as the sheep would be in like a shot.

I continued pondering as I crept quietly across the paddock. I need not have worried, for Breeze was in full view, filling her mouth full of munch. Intermittently she directed her eyes upwards towards my anorak now billowing in the soft wind, then resumed her nonchalant nibbling. My puzzlement gave way to speculation, and then the penny dropped. Breeze's relaxed manner was solely attributed to the Supervisory Scarecrow, wearing my reassuring scent.

As the sun dropped over the horizon, it was time to put out the welcome mat for Breeze. The thought alone was Emma's cue, she was up and at the ready before we had moved an inch.

The air was heavy with the pungent smell of May blossom, and resounded with the clatter of wood pigeons settling in the trees, as we all crossed the paddock. Breeze's 'powder puff' bum was clearly visible. Roe do not have a tail, unlike other deer species, but an anal circle of white hair. When disturbed, this puffs up, acting as an alarm signal, presumably to other roe in the vicinity.

'Oh, the effervescence of youth,' Bernard spoke wistfully, as Breeze jumped several feet into the air with joy.

His words lingered on as I laboured through my evening's programme of cutting out. 'Only a few more...' I kept telling my protesting aching back '... and that's it forever!' Deliverance was at hand, for tomorrow a 'cutter-out' was to join our team. I was suddenly filled with renewed strength at the prospect.

The following days went like clockwork. Having worn my anorak for a goodly part of the morning, letting it absorb my scent, this now shrouded the fence post, comforting Breeze in her afternoon's foraging, and affording me a worry-free period in which to prepare for our forthcoming craft fair. As this was an eight-day event, masses of stock would be needed, and with repeat orders flowing in from our outlets, we needed every spare minute. As the evening shadows gathered, we had come to rely on Breeze's silhouette waiting faithfully under the anorak.

The prophetic tea-leaves symbolised doom and gloom on that fateful Friday, the day prior to the week-long craft fair. The morning sky was sombre, the air hung oppressively heavy, and by late afternoon a storm was brewing in more ways than one. With my car crammed to capacity with crafts, I left to set up my stall, yelling an instruction to Bernard from half-way down the lane to take the washing off the line before the rain fell. His thumbs-up sign confirmed my afterthought.

Unlike most fairs I wasn't needed to man my stall The village hall committee provided this service, deducting a percentage off all sales for this privilege. This arrangement suited me admirably as it didn't tie me down; I needed only to return at the fair's culmination to collect unsold stock and monies due.

I half expected the street to be dancing with rain as I left the village hall. Instead, the wind blew me to my car.

The drive home, full of twists and bends, was hair-raising enough without sidewinds buffeting me on to the wrong side of the road. I decelerated to gain more control and none too soon, as a deer suddenly stepped out into the road, causing me to jam on my anchors. The young buck remained motionless for a few seconds, then in one graceful fluid leap he disappeared into a thicket. I pulled the car on to the grass verge, switched off the engine, and watched him emerge from the trees.

He began wandering aimlessly across an expanse of scrubland, which until recently had been covered with trees. Now it had been cleared, cut down by chain-saws, like so many of the Border forests. Even now, away in the distance, I could hear the mournful, mechanical destruction of yet another valuable animal habitat.

Continuing to watch the banished yearling until he was almost invisible, I wondered how long he would take before finding an unclaimed territory to call his own. For all my mind was on the buck, my heart was whispering Breeze. All the way home the image stayed with me.

Were it not for the sheets, like full-blown sails, whipping up against my windscreen, I would have noticed Bernard's car was not in the yard. Having wrestled to de-peg the remaining washing twisted around the clothes line, I chased across the paddock pouncing on the escapees, bemoaning the fact that Bernard had tempted providence leaving the washing out in such a high wind. The one that got away, now blowing around Mitch's field, struck a wrong chord. There was something I couldn't quite put my finger on, and with my arms full of writhing washing, I wasn't about to go in pursuit of it anyway.

Come to think of it, where *was* Bernard? He must have heard my car, I told myself. Charging through the out-house, ready to give him the receiving end of my tongue, I fumbled for the back-door handle and pressed down on it venomously.

I dropped the pile of washing as if it had suddenly stung me. The huge note taped to the kitchen cabinet said, 'Time 5.00 p.m., taken Jester to the vet. Anorak blown away, no sign of Breeze.'

This was the last thing I expected. Emptied of breath

I flopped down and let the chair support me as I panicked.

With my body back in working order, or something approaching it, I took to my heels. My imagination ranging over the countless possibilities of what could have happened to Jester, I raced around the forest like a lunatic, screaming myself hoarse in search of Breeze, with nothing in return but the whine of the wind. By the time Bernard arrived home, my emotions were out of control.

My tear-drenched face confronted Bernard. Before my distorted garble had time to break through the tightness in my throat, he was hugging me, kissing the top of my head, placating me ... 'Ssh ... ssh ... ssh ... there now ... there now.'

'Oh God, Bernard, Breeze has gone, now Jester's ill, I'm ... I'm cracking up.'

'Calm down, Marie, Jester isn't ill, she ...'

"Then why did you rush her to the vet?"

My annoying habit of interrupting people's flow of words, didn't go unnoticed.

'Let me finish. She collapsed, but there is nothing wrong with her heart, it's as sound as a bell ... She is just an old lady now, her galloping days are over, but her head doesn't know it.'

Although Bernard's voice was level, I knew he achieved his calm only with difficulty. Bernard couldn't display favouritism, he loved each of them, but of all his girls, Jester was special.

Whilst loving me with all her heart, Jester had never given of her soul, a fact about which I had never deluded myself. This privilege is something which happens without expectations or demands, and undeniably it was Bernard that was the be-all and end-all of her life.

The metallic scraping sound of Jester's door-opening attempts broke into my thoughts. Bernard lifted up the car boot, leaned forward and clasped his hands under her tummy. With a firm but gentle lift, he hoisted her to her feet. 'Come and see Mammy then, sweetheart.'

Almost as soon as Jester's rickety old legs touched the floor, the gusting wind sent her sideways as though she were weightless. Bernard supported the front of her chest in the crook of his arm, and frog-marched her through the yard.

'There's my big girl, come on, my Jess, you can do it.' The underlying tone of Bernard's words of encouragement told me what I already knew ... what I didn't want to know ... then, as my mind released the memory of darling Joy's decline, it was like pouring salt into an open wound. I was utterly helpless to stop the kaleidoscopic images, and the tears spilled from beneath my closed eyelids.

The wind had changed direction when Emma, Sheba and I set out on our last walk of the day. I felt a twinge of guilt at leaving Jester behind, but her obvious inability now to withstand strong winds dictated she stay indoors. With Bernard beside her stroking her tummy, whilst she engaged in a spot of ceiling watching, I doubted she cared two hoots anyway.

Emma's barks at something moving across the field raised my hopes momentarily, only to have them dashed seconds later by the sight of my wind-assisted anorak hurtling home. Leading myself up the garden path, I visualised Breeze running after it, but it just wasn't possible. Breeze couldn't jump the fence, and as the hurdle was still closed, she had to be in the wood somewhere. Having secured the anorak well and truly around the fence post again, we played 'hide and seek', without the hiding part, throughout our frustrated forest walk, with none of us a winner.

The wind had abated when I next checked on the lonely scarecrow, and had dropped completely on my third return. This was somewhat of a consolation, for with the light fading fast, along with my hopes, I knew for certain where I would be spending the night.

'You don't ever have to justify your actions to me, Marie. I fully understand, in fact, I would be disappointed if you didn't. Just give me a hug before you go and I'll see you . . . whenever.'

An age ticked by until sunset gave way to sunrise, and whilst dawn is my favourite part of the day, my mind was not on the wonders of nature this particular day break. Bereft of Breeze's presence, the enchantment had gone somehow.

I had almost reached the old oak, my final point of call, for the umpteenth time, when my mind short-circuited. Maybe she could now leap the fence – what if she had? – and suppose she has left forever . . . The chill of reality hit me like a ton of bricks.

Why, when I should have felt elated at Breeze's final accomplishment, did I feel so dejected? How often in my secret moments had I not prayed silently for freedom to call her, saving me the heart-breaking task of turfing her out into an unknown world? Now that my wishes had come true, I felt the gnawing pain of emptiness. Perhaps it was because of the suddenness of her departure – no goodbyes, no reflective moments, just *gone*.

In the days that followed, waves of longing for Breeze swept over me. And every time I looked at Emma I cried. Her misery was mirrored in her doleful eyes, and in the sighs when she knew for certain she was sleeping alone. Not that she really slept, she just dozed, with her ears permanently straining. The slightest sound would send her on a fool's errand, like the night the mouse came to visit and she shot into the kitchen as the patter of tiny feet crossed the kitchen floor on its foraging expedition.

Day and night, my crafts became my salvation. I couldn't sleep, as my mind just wouldn't settle, churning over memories. My restless spirit kept questioning as well. Why such a sudden desertion? Why on that particular day? The forest was Breeze's domain, so why leave it at all? All those months ago, I had read that a female yearling fawn is reluctant to leave its birth territory, and after being ousted, will repeatedly return; she will eventually learn to avoid her mother, while continuing to live in her vicinity, so why would Breeze leave of her own volition? Something didn't quite gel, and it constantly stalked the surface of my muddled mind.

An endless relay of coffees washing down a panacea of fizzies and a total absence of sleep, had me feeling that I was defying gravity. Life seemed unreal somehow. Every waking minute had revolved around Breeze, and now that she had gone, I felt drained of my life-force, as if part of me had gone with her. Were it not for Bernard's constant bolstering of my emotions, I might well have given in to my misery.

I lay counting the raindrops beating rhythmically against the window as I tried to summon sleep for the third successive night. Drumming, drumming, they fell faster and faster, then, as the heavens opened, my conscious mind finally slipped, carrying me along in the flow. It was a strange dream, I wasn't in it, it hadn't any thread and the scenes were motionless. It was purely a series of flashes, like watching a slide projector showing the same three stills over and over again. A river, a raging torrent, then three snarling dogs in a stance that indicated they were holding something at bay.

The final frame flashed. The lace curtain hung suspended in air and through it was a misty shape with the face of a fawn, knee deep in a quagmire. I felt a scream building up in my throat, and a hand stroking my hair, then a voice broke through and my dream dissipated.

'Good God, Marie, what a blood-curdling scream! I thought we were being murdered in our beds!'

'Oh, Bernard, Bernard, I had this horrible dream, it was so vivid, Breeze was sinking.'

'There, there,' Bernard wiped the tissue over my perspiration-soaked forehead, 'it wasn't real, it was only a dream.'

Although my head nodded in agreement, my heart was disobediently murmuring something to the contrary.

I firmly believe it was this push from fate that took us to the river on our morning walk.

In consideration of Jester's condition, bless her, our walks were now confined to more even ground. We had only intended to saunter down the lane as far as George and Essie's gate. The rain had finally stopped, the air was still, and the pleasant penetrating sunshine seemed to give Jester a new lease of life. It was her desire to sniff further afield which led us down to Mitch's meadow in the first place.

As I sighted George at the bottom of his field, accompanied by his three collies, examining his flooded land, my eyes were uncontrollably drawn as though by an invisible magnet to the rushing river. I felt a sudden headiness, then a visionary flash, as a feeling of déjà vu washed over me. As Bernard and George locked into conversation, a perception stronger than intuition told me what he was saying.

Two days ago George had caught sight of a yearling on the other side of the river bordering his land. The water level then was low enough for her to wade across, but her repeated attempts to do so were thwarted by his dogs. His last glimpse of her was yesterday evening, standing watching the swollen river.

George took a puff of his pipe but, as usual, its aroma didn't drift. Considering the bowl was filled with baccy, it might as well have been stuffed with hay, for although the pipe was constantly in George's mouth, we had yet to see him light it!

'It mightn't have been your bloody deer, mind,' George said, wagging his pipe at me with a wink in Bernard's direction, betraying his wry sense of humour.

'Oh, it was Breeze all right – make no mistake about it,' Bernard replied emphatically, his brow furrowing into a frown, 'but for the love of me I can't understand how she got herself in such a predicament in the first place.'

'You know, it's my guess, Bernard, she pursued my "runaway" anorak thinking it was me and lost her bearings.'

'However, Marie, she certainly can't cross this avalanche of water.'

Jester's interest in anything much beyond her nose having waned, she gave Bernard an 'I'm bored now' nudge, and so we returned home, but we were determined that Emma would have a sleeping partner that night.

Dusk hung heavily as I hurried down the lane again, alone. Sheba and Emma were still exhausted from their last search, which had been all of an hour ago. I had set my mind on somehow crossing the river at its shallowest point and working my way upstream. I thrust three sore-throat pastilles into my mouth in an effort to oil my seized voice-box, and broke into a trot. Half-way down the lane I decided to take a short cut across the field when something caught my eye.

'It's probably only a sheep,' I thought, as I slowed down to take a second look. The faint shape moving towards me suddenly became visible, and every molecule of my being sang with pure joy.

Sleep didn't claim us that night. Exultation did!

Emma's remarkable ability to express a variety of moods in behavioural gestures is surpassed only by her eyes, and there was no mistaking the maternal tenderness in her look as she 'groomed' her baby. The eye communication was mutual, as Breeze repeatedly upturned her face, eager for more. Finally exhaustion overcame her.

Watching, loving her asleep, I thought how fortunate we had been in life's lottery.

18
In the Family Way

The morning saw the beginning of a strange ritual. After my endless leg massage, Breeze then manipulated her body around Bernard's thighs before bonding herself to items of furniture and doors. Having continued this abrasive contact on every surface leading to the paddock, we were left in no doubt that she was scenting her home ground. As only the *buck* marks his chosen territory by the rubbing of a forehead scent gland against bushes and trees, it couldn't be anything other than an identification trail, her marker to home and us.

'This way, Breeze...' The words were hardly out of my mouth before Breeze was over the fence, leaving me holding the hurdle open, and a somewhat bewildered Emma waiting to usher her in.

Flaming June turned into a scorching July, with Breeze hopping in and out of the forest, sometimes for minutes, sometimes hours. With our front door open 'all hours', she was now free to roam wherever her fancy took her, as and when she wanted to. Strangely enough, she was always home just after dark, the rattle of the dog biscuit bowl and crunching announcing her return. Only then was the key turned in the lock.

BREEZE: WAIF OF THE WILD

Breeze had never been subjected to our rules, more the reverse, and even now she was a 'free spirit'. The fact that she still chose to spend time with us was solely her decision.

We had deluded ourselves into thinking we were prepared for the inevitable. Unlike Joy, Jester never knew that her time had come, she was so used to Daddy lifting her up when her legs gave way. The sun-bed was her destination at her final sinking, so fitting somehow. As with Joy, the vet's gentle and tender touch allowed her to die with dignity. Jester was out to the world, reaching for a sunbeam, when Joy called her to play in eternal sunshine.

Her loss provoked diverse emotions: sorrow, outrage, guilt and depression. Jester was more than a loved member of our family, she was an extension of Bernard. The heart can be given, can be broken, repaired and given again, but the soul, once given, is for eternity. I knew that void in Bernard would never be filled. Jester was physically laid to rest alongside Joy, but she was buried forever in Bernard's heart.

The hens clucked enthusiastically as I scattered their morning corn. 'Ooh, six eggs today, thank you, chuckies.' Carefully cradling the eggs in the crook of my arm, I turned to leave the hen house.

'Peep, peep, peep . . .' The terrifying note cut straight through my heart, eclipsing all feeling, the eggs smashing to the floor.

'Bernard! Bernard!'

We reached the front garden simultaneously, just as Breeze flew out of the forest with a leap that would have done an Olympic pole-vaulter proud. With split-second timing she was across the lane and into her garden sanctuary, leaving her pursuer in no doubt that she was frightened out of her wits.

At first glance we thought it was Mr Fox, due to the flaming red coat. It was his coronet that assured us

Breeze had a suitor. Alas, his ardour was soon cooled by the sight of Breeze's big barking bodyguard. We watched the frustrated young buck's speedy retreat. We knew full well that he would be back though, for it was now late July, time for him to fulfil his biological urge. It was also time for Breeze's departure from her worry-free existence of fawn-hood, for she was now coming into 'season'. Unlike other British deer which mate in the autumn, the roe's sexual activity begins in mid July and continues until the second week in August, the rut of the yearling doe being the earliest.

'Here comes Mr Twonks, Marie,' Bernard beckoned me to the kitchen window. We watched the handsome young buck advance on the garden yet again. 'I'll give him ten out of ten for persistence. I wonder if he . . .' Bernard's thoughts were cut off as a car sped past and he was lost from sight, foiled yet again.

'Hey, Twiddly Twonks,' Bernard said, wagging his finger at Breeze, 'your boyfriend came to call again. Aren't you interested?' Whether he didn't strike her fancy, or she was playing hard to get, Breeze's 'I couldn't care less' attitude as she continued chewing her cud, sandwiched between her two chaperones, brought a smile to my face.

'You'll stay an old maid if you don't get your act together, Breeze.' Bernard's delivery, like some stern Victorian father, turned my smile into a fit of giggles.

We were expecting Breeze to join us on our usual evening walk down the lane, but she had taken herself off on her own. We looked across by force of habit as we passed Breeze's leaping 'in and out' part of the forest, and just inside the perimeter fence a pair of ears twitched under an umbrella of bracken. Normally she jumped out when we passed by, and honoured us with her presence. Tonight we qualified for no more than a cursory glance.

BREEZE: WAIF OF THE WILD

There was a simple explanation. She was not alone. Mr and Mrs Twonks were lying side by side. We watched through our binoculars, totally fascinated by their circular game of 'catch me if you can', with the young buck doing the chasing whilst Breeze uttered soft enticing peeps. Darkness finally called a halt to our activity, but certainly not theirs. We knew who Breeze would be spending the night with, and it certainly wouldn't be Emma.

It was noon of the third day before Baba's eyes lit up again, and being perfectly honest, mine too. I had actually given more than a passing thought to her beau having led her to pastures new. We had both agreed, once Breeze could negotiate fences, that the woods would become out of bounds to us humans. This was Breeze's natural habitat, not ours, and we weren't about to venture in simply out of curiosity.

'You needn't look all coy at me, you little object of desire,' Bernard spoke teasingly. Breeze rubbed up against him, assuming an expression of blameless innocence. 'We saw you leading him on.' I swear Breeze understood his words, for her expression was coyly bashful as she wiped her face on my thigh.

Mr Twonks never called again, nor did any other wooer, and the letter that arrived from our friendly vet explained why. 'Thought the enclosed might be useful, hope to hear the good news next year sometime.' The fact sheet on 'Mating and Reproduction of Roe Deer' was a revelation. Bernard carried on reading aloud, 'Once fertilised at her first successful mating, the doe immediately goes out of oestrus [season] and is of no interest to any virile buck.' He paused, speechless.

I jumped out of my chair, flung my arms around him and we waltzed around the kitchen singing in joyous unison.

'Breeze is in the family way, family way, family way, Breeze is in the family way – our little Breeze.'

In the Family Way

Roe have an exceptionally long gestation period of 294 days, in comparison with other deer, whose gestation periods vary between 176 and 245 days. They also have a very unusual reproductive cycle following fertilisation, for whilst most animals grow continuously inside the mother, the roe fawn does not. After the first few days of the embryo growing normally, it is checked as it enters the uterus and lies dormant for five months until the advent of the new year, when it begins normal growth again, continuing for a further five months until birth. This delayed implantation of the ovum occurs in no other species of deer, nor any other cloven-hoofed animal.

Prior to the Ice Age, roe could give birth at any time of the year, but with climatic changes, the young would not survive in freezing weather. It's hard enough for the does themselves. Breeze was fortunate, as she had never experienced pangs of hunger. She shone with condition, and undoubtedly when next May saw Breeze a mum, her fawns would also be in fine fettle.

Our moment of celebration was to be short lived, for within the hour a Forestry Commission vehicle drew up outside the house. A stranger and another, whom I recognised as Billy, got out, the girls raised the alarm, and Breeze took off. As is so often the case, the telephone began its untimely ringing. Bernard made for the lounge in answer, whilst I dashed outside to quieten the vigilance committee and greet the wary callers.

Billy, the 'owl man' as he is known locally because of his rehabilitation of injured owls, is employed on the conservation side of the Forestry Commission. Having first introduced me to the ranger, Billy then enquired about our barn owl. He had made and erected a special nesting box in the rafters of the barn last autumn in the hopes the owl would rear young, and was saddened to learn that, as yet, it hadn't.

'We've come to site a couple more nest-boxes amongst

the perimeter trees,' Billy stated, reaching into the van and extracting the small boxes. 'We'll not be very long.'

As they went about their business, Bernard entered the yard carrying two cardboard boxes which he proceeded to put into the car.

'Billy and his mate have gone to put up bird-boxes, Bernard. Where are you going to?'

'I'm just nipping down to Bessiestown, pet, as Margaret has run out of hedgehogs. I'll be back in half an hour.'

Breeze was more curious than alarmed by the men and what they were doing. She stood stretching her neck to the fullest extent, watching only yards away as they removed the bottom branches from a tree and erected a small nesting box.

The moment they had made their exit over the forest fence, Miss Nosy Parker trotted up to the pile of branches and gave it the third degree.

It was the ranger who spoke first. 'Is that your pet deer?'

'She isn't a pet,' I retorted. 'She is a wild creature that chooses to live with us.'

His eyebrows shot up to his hairline. 'That's somewhat unusual, isn't it? I've never heard of a deer living in a house full of dogs.'

'Well, this is where she was raised,' I smiled. 'She was only the size of a bag of sugar when she was orphaned, and we are trying to return her back to the wild, but it takes time... You will know this only too well,' I added, turning to Billy, thinking of his owls.

The ranger's next words were insensitively to the point. 'I doubt very much she will survive in the wild, as she hasn't any fear of man – she's a sitting target when it comes to culling.'

I was struck by a shock-wave hitting every nerve in my body.

Billy, seeing the colour drain from my face, quickly

In the Family Way

jumped in and addressed the ranger. 'They could mark her in some way, say an ear tag, and you could let the other rangers know?'

'Aye, that'll be the only way we would recognise her, Billy. She could be branded, but it would need to be on both sides, maybe even a collar. No, I think tagging would be the best, it's more visible.'

By now my mind was screaming. 'Breeze is only one year old, she has seven or eight years ahead of her.'

How could anyone think of exterminating an innocent whose only offence was in having survived a year, being fruitful and hoping to multiply. She had battled against all odds to enjoy this lifetime of happiness, yet this very point was at issue! This wasn't selective control, this was wholesale eradication! Why had I thought only the aged, sick and weak were culled! Who else knew age was immaterial? Who the hell cared! I suddenly hated the world of humans.

A new thought now crowded my mind. Futility. The future I had envisaged for Breeze was obliterated by the devastating reality I could never have foreseen. The rapid pounding of my heart began again. On the pretext of having to rescue my baking from the oven, I turned hastily away, and the vehicle drove off.

I had always found yoga a reliable 'unwinder' in the past, but even this let me down now. My mind was just too cluttered to meditate. I pulled my knees up to my chest and, wrapping my arms around them, rocked back and forth giving vent to my emotions in the only way I knew how, by releasing floods of tears.

I met Bernard's return with red-rimmed eyes and a frog in my throat.

'Do you know what the ranger's just said?'

Within ten minutes of having told Bernard of the situation, I was on my way to consult Ken. Guilt nagged at me as I drove to the vet's. The more I thought about it, the more incensed I became, and my inner voice asked a horde of questions. 'Why should a free spirit be shackled by something as degrading as a stock identity disc, it's an imposition, an insult.' 'Yes, yes, I know,' I answered myself, 'but it's no good living in sentiment, you must face reality.' 'Perhaps the vet will come up with an alternative solution.'

I was full of optimism as I waited in the surgery for Ken's advice, but in the doldrums when I left.

'Freeze branding', a method used to identify horses, was the gentle answer, being totally painless and safe in the hands of a skilled farrier, which I was not, and as Breeze would not tolerate being 'handled' by anyone other than myself, this was out. A luminous collar, my suggestion, was not advocated by Ken, as she could find herself hooked on a barbed-wire fence. We were left with tagging as the only possibility.

Ken was candid. 'Yes, she will feel it, and in all probability she will bolt, it's an animal's natural reaction to pain. The most important factor is that you pierce the exact spot, because there is a main vein running through a deer's ear and if you rupture this you're going to have haemorrhaging, with no way of stopping it.' Ken picked up a pencil and began drawing. 'The vein is situated here, and this is where you must place the tag.'

My eyes refused to focus on his diagram for I knew I would not, could not, knowingly inflict pain on any creature.

'I'll do it for you.' I listened with a grave expression to Mitch's obliging offer. 'It's second nature to me, Marie, I've been tagging sheep for years and she won't even know I've done it.'

'Yeah, literally, Mitch,' Bernard replied, his flash of wit sending Mitch into a hearty laugh. It was strange how Breeze regarded him as 'The Invisible Man', even his greetings went unheeded as she brushed past him ignoring his presence.

'Think about it, Marie, I'll not be long.'

In the short time it took Mitch to return with the ear-tagging equipment my mind was made up. Although Mitch's intentions were good, my answer was 'Thanks, Mitch, but *no* thanks.'

19
Marking Time

A cloud of dread hung permanently in the back of my mind as I watched Breeze leaping into the forest with increasing regularity. As summer continued, my concern was transformed into obsession.

We were washing the car when Mitch appeared, waving a leather harness. 'Bernard, can you spare me a hand to catch the tups?' he shouted, but then Mitch very rarely spoke softly.

'Sure, just give me five minutes to hose this down and I'll be with you.'

'It's mating time again, is it?' I enquired.

'Yeh, lucky old tups!' His mouth twisted into a smile.

I watched Mitch insert the crayon into the harness and marvelled at the ingenuity of such a simple device. As the tup covers the ewe, the wax crayon leaves its mark, and the number is noted, giving an estimate as to which ewe would lamb. This information apart, it's also a sure-fire indicator that the tup is doing his stuff, as each one has a different coloured crayon.

I had a sudden flash of inspiration. 'Would I be able to draw a collar on Breeze with that crayon, Mitch?'

'I don't see why not, but it isn't permanent, mind, it will wear off.'

'Well, I'll just keep on re-marking her. Where will I buy a crayon from, Mitch?'

Obliging as ever, he held up a selection. 'Which colour would you like?'

The following week, both Breeze and Emma looked like Red Indians on the warpath, with items of furniture, doors, plus our clothing, all daubed with orange-red wax.

The tups, which had usually chased Breeze when she ventured too close, were now fleeing from the warrior. In fact, Coldside became an area to avoid as 'Big Chief Twiddly Twonks' would fly over the fence the moment she heard sheep munching, then, having evicted all trespassers, would stand victorious, raising her foot and stamping it down with supreme confidence.

Friday was drab and damp. Thankfully it didn't deter Breeze from her tour of the forest, and so allowed me to carry out my set task for the day. It was four o'clock before I'd finished shampooing all the carpets, and still the rain was falling, albeit moderately.

Midnight saw puddles on the piece of tarpaulin placed over the carpet in the front hall, but no sign of Breeze. At two in the morning, conceding Breeze had missed the last bus home, we kissed the girls 'night night' and called it a day.

Overnight, the open door and welcoming light had attracted only midges and moths, contrary to Emma's expectations.

'Don't look so sad, Baba, she'll be back soon,' I repeated, all of a dreary wet Saturday and Sunday. By this time it was more of a plea from my heart than an assurance to Emma.

The sun streamed out of a sapphire sky as I set off early on Monday morning. The Priory's annual craft fair had attracted an extraordinary number of visitors, and my stall was looking 'depleted' according to the organiser's telephone call.

The country roads as usual were traffic free. Unlike driving in the city, when one's eyes are always on the vehicle in front and behind, I was able to look around me. The conifer plantation to my left looked as if the heart had been taken out of it and, rounding a bend, a mountain of stacked tree trunks, alongside a warning sign 'Timber Felling in Progress!', explained why. A magnificent roe buck, his antlers held high, in search of a doe coming into late season no doubt, was crossing the deforested area, and his vulnerability caused a pang of pity.

It also caused me some alarm, for it was still the open season – the legal shooting period – for bucks. In England this lasts from April 1st to October 31st; it is November 1st to February 28th for does. (In Scotland – not so very far away – the open season runs from April 1st to October 20th for bucks, from October 21st to March 31st for does.)

'Hide, hide, go back to the shadows, don't you know it's still open season? To show yourself is to court disaster!' I yelled.

My thoughts switched to Breeze. Although it was still close season for does, when they are protected, this was drawing to an end. Thank goodness she was now identifiable, I consoled myself. 'But for how long,' Thomas the Doubter asked. 'Suppose the rain was to remove all traces whilst she was off on her wanderings, what then?' 'Be logical,' I should have replied, but alas, not falling into this category, I'd emptied a box of tissues on my streaming eyes by the time I reached home.

Never was I more relieved to see my freshly cleaned carpet besmirched with deer droppings. Breeze wasn't about, nor was anyone else for that matter.

'Charming,' I said aloud. 'All gone for a walk leaving me to clean up.' I was still on my knees with the brush

BREEZE: WAIF OF THE WILD

and shovel when Bernard returned.

'What on earth are you doing on the floor, Marie? Oh, I see the happy wanderer's back then.' Bernard indicated the shovel, pinching his nose between his finger and thumb.

'Isn't she with you, Bernard?'

'No, I haven't seen her, she obviously paid us a visit as we were out on our walk and kindly left her calling card.'

'Where are the girls, Bernard?'

'Need you ask. The moment I opened the gate they made straight for their sun-beds.'

'Dispose of this outside, Bernard, will you, whilst I get the disinfectant.' As I thrust the shovel at him the fumes lodged in his nasal passages and the spluttering began. Pegging his nose again, he dashed out of the door.

'Don't bother with the disinfectant, Marie, Twonkers is coming up the lane!'

'I see your collar has washed off then. I'll have to draw another one, eh, Breeze?' I said, feigning nonchalance as she wiped her nose across my skirt.

'That didn't stay on long, did it, Bernard?' He remained silent. I didn't add what was really in my mind. I had no need to anyway, for Bernard had read it already.

'Coffee, pet?' I asked, changing the subject. Again, Bernard didn't answer, his eyes were fixed on Breeze. He was still deep in thought as he sipped his black coffee.

'Model aeroplanes,' he mused aloud.

'Model what?' I looked at him curiously.

'I've had an idea, Marie. Will you bring me the Yellow Pages, please?'

The hobbies shop had exactly the non-toxic fluorescent paint we required, guaranteed to be waterproof, unlike Breeze's last ineffective marker. As she settled down with Emma for the night the tips of her ears were painstakingly transformed into beacons.

It had taken me over an hour as each brush stroke, a mere whisper of a touch, brought about an irritable head shake. My patience was rewarded, though, as the results were marvellous. Two conspicuous luminous lanterns now stood out in the dense vegetation. Who could fail to see her! I informed the ranger accordingly.

Breeze did precious little of anything other than laze about for the rest of the long extended golden summer, stuffing herself with biscuits most of the day, in preference to a forest picnic. However, she always took herself off at about seven in the evening, when the air had cooled, returning around eleven. We strongly suspected she foraged at dawn, not that we heard her leave, but by the absence of droppings around the house.

Inevitably, autumn nipped the air, and everything took a nose-dive. Little wonder that this season is referred to as 'the fall'. As showers of shivering leaves and plummeting cones littered the woodland floor, our hectic craft season thankfully quietened down. We were certainly in need of a 'breather' to recharge our batteries before the Christmas orders came flooding in again. With six seasonal craft fairs booked, this was a period of inactivity we were more than looking forward to

No sooner had we dropped down a gear than Breeze went into overdrive. She was in and out like a yo-yo. At first, gone all day, followed by just a night out, then her away days stretched into night, and her night overlapped with day until I didn't know who was coming or going, her or us!

I had come to dread the nights of a full moon, purely because of Breeze's uncontrollable urge to wander away for days on end. What with foxhunts during these periods and the threat of rifle shots, her sudden desertions were playing havoc with all of our nervous systems.

So many times I lay in the quiet of the morning before

sunrise, my ears straining, listening for Breeze's footfall on the gravel. I wanted her home with us, at the same time praying that her wild side would take over. This dualistic thinking was driving me round the bend.

When Breeze chose to return, smelling of pine, stuck with conifer needles, chunks of her coat missing and thinner, she was always on the point of exhaustion as if she had travelled miles. My unbalanced imagination had her being chased from her familiar surroundings, lost and spending days finding her way home.

Problems have a way of entering our household. In fact, I'm convinced they permanently live here, lurking in some dark corner, ready to pounce on us when we least expect them! At first it was just the odd hair that fell out, then a few more, and as the ear scratching began, luminous paint and all came off, exposing weeping sores.

'Allergic eczema, I'd say, in view of the facts,' said the vet. For days poor little Breeze had antiseptic-coated ears. The antibiotics she never knew about, as they were craftily concealed inside a 'custard cream' biscuit. We were now back to square one – no identification tag!

It was the craft item I was making that caught Bernard's attention. 'Excuse me, Marie, how are you going to get that wide elastic to fasten?'

'I sew a piece of this "wonder webbing" to each end, and when you press it together it bonds, then you merely pull it apart to open it, like so.' I could almost hear Bernard's mind ticking over as I completed my demonstration.

'Couldn't you make Breeze a collar like that then?'

'Well, I suppose I could, but you can't buy fluorescent elastic, Bernard.'

'Oh, what a pity, it would have been ideal as it's so flexible.'

Bernard's brain wave had me putting my thinking cap on, and I found the solution in a bicycle shop. Having purchased a luminous arm band, I stitched a strip of the fabric on to some wide elastic, sewed the wonder webbing to each end, and Breeze now wore a glowing red necklace for every ranger to see.

It was late afternoon on November 16th, when the first flakes fell, disproving our belief that it was 'too cold for snow'. With our craft fairs cancelled, our account with the greengrocer renewed, and red bows tied on to the snow-draped trees, we settled into our early winter.

The snow came and went throughout December, but not Breeze, she remained constant. She honoured us with her company on our walks, enjoyed the odd trot around the garden, and spent her nights by the fireside, next to Emma. How we cherished our cosy evenings, watching her rapid eye movements and jerking delicate limbs as she dreamed.

However, depression set in on December 20th, for Breeze had been missing for three days.

The north-easterly arrived, changing the snow showers to a raging blizzard, and still the front door remained ajar. As fast as we shovelled the accumulated snow out of the lobby, it blew in again. Higher and higher it mounted, until going up the stairs was an impossibility, for we could no longer see them. The lounge became our sleeping quarters. Not that we slept much, if at all, as the nights were spent flinging bucketfuls of coal on to the fire and sipping steaming hot cocoa in an effort to stave off the draught.

On December 24th, in the snow-carpeted room off the lobby, the Christmas tree stood, devoid of decorations, neither of us being able to dredge up any Christmas spirit. Breeze's biscuit box and vegetable bucket, full to the brim, lay untouched for what seemed like seventy days, not seven. Bereft of its heartbeat, Coldside felt dead and the silence was screaming.

Christmas Day didn't dawn, the sky was too busy still shedding its load. Never had I ever experienced so much snow. Fall after fall cascaded from overloaded branches, doing nothing for my ragged nerves at all, as each earth-shaking thud sent the girls into a bout of barking and kept them forever on their guard.

'Oh no, not again,' Bernard said wearily as the lounge door rattled. 'I've only just finished clearing the last avalanche from the lobby.'

It wasn't what Bernard said that made me palpitate and slam down the iron, but what I did *not* hear. No 'woof'! Emma suddenly sat bolt upright, stared intently at the door, and with her eyes widening by the second, she began to tremble, then grunt.

'It isn't. It can't be. It *is*,' Bernard shouted aloud, overturning his chair. A bustle of excitement greeted Breeze's entrance as if she was royalty.

Of all the Christmas Days past and those yet to come, I knew none would ever be as divine as this one. As we unwrapped the gifts we had received, the one that really gladdened our hearts was the one we hadn't planned. Oh, how regenerative happiness is.

20
Dawn and Dusk

The New Year saw a perceptible change in Breeze. She began to seek privacy more, cuddling up to Emma for a short time, but more often than not she chose to be in the utility room curled up on the floor. It was her breathing, rather too rapid, coupled with a marked lethargy, that began my concern, and with every puff and pant my solicitude deepened.

'Here we go, pressing your panic button again, Marie! She's obviously too hot! How would you feel wearing a two-inch-thick fur coat, a fire roaring up the chimney and every radiator on in the house?'

Bernard allayed my alarm in this respect, but when her urine turned a rusty red, I nervously conferred with the vet.

'No, she doesn't appear to be in pain, just panting a lot. Yes, her motions are perfectly normal. Yes, she is eating well, *very* well.'

'Frankly, I don't think you have anything to worry about, Mrs Kelly. Embryotic development will now have taken place, with her young now being attached by a placenta, and her coloured urine is certainly due to something she has eaten, rather like us eating beetroot, not an indication that anything is amiss. The central heating is probably the cause of her panting. There really is no need for concern. I look forward to hearing the good news around the end of May.'

For weeks I soared as gossamer in a cloudless sky, until February 19th brought me down to earth.

'I'm afraid she has had a slight stroke, and it's affecting her back legs. Fortunately, being a lighter dog than Jester, she could carry on for months.'

I had difficulty in meeting the vet's eyes. Whoever said, 'To be forewarned is to be forearmed,' got it all wrong. Nothing was going to soften this blow.

Sheba, who had become Bernard's constant companion after Jester's death, was now in the departure lounge of life and as with Jester, we were faced with the insufferable uncertainty of *when*. The pain of the past echoed in the present, and haunted me all throughout the night. Bernard was occupying the spare room, lost in his own private thoughts, and from the amount of nose blowing I could hear, he wasn't doing much sleeping either.

I felt a sinking sensation every time I looked at Sheba, and with Breeze off on yet another unannounced vacation, I spent most of the following day wrapped in gloom.

'Whenever I am troubled, Marie, I focus all my attention on one task. I find doing a good wash therapeutic. I strip all the beds, blankets and all. Watching them blowing in the wind kind of blows your cobwebs away. Cleanliness is next to godliness, so the saying goes.' It was after this telephone call from my mother, offering me her cure for the blues, that my compulsive behaviour began.

The time of year was hardly conducive to a line full of washing, so I set about an exceptionally early spring clean. Using the excuse that the house was appallingly neglected, I gradually wore myself to a frazzle.

Everything that could be cleaned, even if it didn't need to be, was. Anything to take my mind off Sheba.

By the middle of the week, having gone over it again and again, feeling drained and totally exhausted, I still managed to trick my mind that work was good for the soul, until finally I ceased to function as a person. As I continued scrubbing around the clock like an automaton, Bernard's concern deepened.

'I think you really must ease off on this cleaning ritual, Marie.'

'I can't, I must keep busy, Bernard, it keeps my mind under control.'

'Do you know what happens to a guitar string when you tighten it and keep tightening until it becomes intolerably taut, then you tighten it one more time, Marie?'

'You don't have to be cryptic, Bernard, say what you mean,' I snapped, slapping the threadbare floor-cloth on to the lino again.

He shook his head, cleared his throat and took a deep breath. 'Marie, you are wearing yourself out, and not only physically . . .'

Bernard seemed to be whispering, as a strange spinning sensation came over me. The clock was beginning its noon chimes when everything faded out.

I knew nothing more until my prince kissed me awake. 'Welcome back, Sleeping Beauty, your breakfast is ready.'

Along with the liberation from my addictive energy, I'd had a shift of mood during this long, serene sleep. As I spent the day doing nothing, I hummed to myself.

April arrived unbelievably mild, and so did Mitch, as the lambing was favoured for once.

'Grand morning . . .' I understood, but whatever was said in between this and . . . 'Is Breeze back the noo?' I could only hazard a guess, for when his words were well laced with his Scottish accent, even Bernard had difficulty in grasping their meaning. I shook my head in the negative and continued cramming peanuts into the swinging coconut shells.

'It's been two weeks now, hasn't it?'

'Seventeen days, as of today, Mitch,' I answered, feigning nonchalance.

'I reckon she'll be gone for good, don't you think?'

'Oh, no, she *will* be back, Mitch!'

He raised a smile, shook his head, turned the ignition key, and started up the tractor. 'She has to leave some time, Marie, she can't stay forever.'

'I realise this, Mitch, but I'll know when the time comes, and it's not *now*,' I yelled as he drove off.

I had a very strong feeling of second sight as we returned from our afternoon walk. As Emma rushed towards the utility room, her eyes brilliant with anticipation, I knew my intuitive sense had proved me right again. We heard the girls' tails thumping with pleasure, and peeped in. Resting on the settee, her forelegs stretched luxuriously out, her head in between, was Breeze. The joy of her return reflected in the happy faces all around.

Our first concern was for her condition, but not wanting to spoil the happy reunion, we allowed her to be fussed over a little.

'Off you go now, Breezie is tired,' Bernard urged. As if understanding, they all trotted out, without so much as a sigh.

Although on the face of it Breeze looked perfectly normal, Bernard examined every inch of her body with concern. Having pronounced her fit and well, but very, very tired, we made our exit, allowing her to recover from her journeying afar, which, as with all her previous away periods, still remained something of a mystery.

And it was with gentle pride and lumps in our throats that we watched Breeze's final emergence into adulthood, throughout this unforgettable spring. With her moult now complete and the rich colour of amber clothing her fullness of figure, our times together lessened as her periods of independence increased.

Clumps of pale primroses and meandering marsh marigolds wallowed in the late April showers, whilst the buds of May perched patiently on the hawthorn hedges awaiting the departure of the contradictory weather. Predictably punctual, swooping up and down and to and fro, the swallows arrived on May 1st, bringing the South African sunshine with them. As it is usual for adult birds to return to their breeding ground, my jubilation soared with them as they took up occupancy of last year's mud cup, high on a ledge in the coal house.

Straw, feathers, moss, twigs and sheep wool were navigated through the air with relentless regularity as the nest-building days of May gathered momentum. By mid May the birds' breeding season had begun in earnest and, according to the 'book', so should the roe deers'. A dog would indicate the nearness of birth several days in advance by tearing up papers or digging a nest, by refusing food and generally acting restless and nervous, but the only changes in Breeze so far were of a physical nature, in as much as she was growing heavier by the day, and a broadening across her loins was evident.

A nagging thought had entered my head and, try as I might, I found myself fighting a rising tide of apprehension.

'Bernard, do you think we should stay up with Breeze overnight?'

'Whatever for, Marie?'

'In case she decides to have her young while we are sleeping.'

Bernard's spontaneous belly laugh raised my hackles. 'I don't think this situation is funny at all!'

'Situation? There isn't one. Come on, be sensible,

Marie, she's not going to give birth in the house with the girls about, all animals seek solitude at this time. Anyhow, if there's one thing I'm certain of, it's that Breeze's young will be born in the wild, so stop all this silly worrying.'

Easier said than done. So convinced was I that one morning I'd come down the stairs and find not one deer but three in the lounge, that I went to bed later, lay awake longer and woke up earlier.

Ironically it was May 25th, her 'given' birthday that saw her prone to swift changes of mood. From quietly sleeping one minute to sprinting around the garden the next, then panting, 'peeping', scraping Emma's sun-bed, flattening it down and herself with it, only to jump up again with a start, left us in no doubt that her reproductive cycle had almost turned full circle. Whatever we had planned for today would have to go by the board.

Whilst the kettle was boiling, I hurriedly made up a couple of sandwiches, poured the hot coffee into the thermos flasks, grabbed my glasses and needlework, threw the lot into a basket and joined the family in the garden. The butterflies weren't only fluttering in my tummy, they were doing somersaults in my heart as I awaited the final countdown like an astronaut. However, I wasn't the one taking off.

I picked up my embroidery, but within seconds had put it down. Bernard looked as if he were attempting a crossword, but judging by the shake of his pen, I doubted his ability to concentrate either. Feeling tears sneaking up on me, I took a deep breath and fixed my gaze on the various shades of green. What a jewel of a tree aspen is, with its silvery green bark studded with black diamonds, and its delicate quivering leaves like a thousand silver wings. Folklore has it that this tree harbours secret grief, having provided the wood for the cross on which Christ was crucified. Even today, with just a hint of a breeze, its leaves were all of a tremble. I felt a cold shiver slip down my spine, and I suddenly felt as one with the aspen.

The trees began to cast shadows, and still we waited. Breeze rotated her neck with the pliancy of a willow as Bernard unclasped her collar.

'I think it's constricting her throat, pet, she is better off without it.' He smiled a soft, sad smile and I said nothing, for hadn't I always known that Bernard had regarded it as an assault on her liberty? Breeze entwined her body into Baba's, dropped her head and succumbed to sleep.

The shadows were just beginning to lengthen when Breeze awoke with a sense of urgency. The waiting was over. We were suddenly locked in a three-way embrace. Breeze was rubbing her face against Bernard's arm, his thigh, kissing my nose, nuzzling my hair. I felt a compelling urge to gather her up in my arms just as I'd done the first day I set eyes on her.

The inner fragility I had felt all day was nothing compared to the anguish that pained me now. I felt the tears start, my shoulders shake and I collapsed in Bernard's arms.

Gently, he kissed the top of my head. 'Hang on, love, please ... please ... let her go in happiness.'

'Oh God, it hurts so, Bernard ...'

'Yes it does, pet ... I know ... I know.'

Her walk across the garden seemed agonisingly slow, then in one majestic leap she was over the hedge, galloping purposefully towards the forest. With one hand in Bernard's, the other clutching my trembling heart, I held my breath in anticipation of her final departing vault. Instead she hesitated, and wheeled around to face the garden. There was no mistaking the message in her lingering look, and I know she heard my inner whisper of, 'and I will always love you too.' Then with ears pricked she soared into the woods in pursuit of the unseen caller whose voice was profoundly more powerful than ours.

Breeze's final departure had such a devastating effect on us all, not least Emma, who never stopped looking behind her on our walks, in the forlorn hope that Breeze was following.

My feelings vacillated between coping very well and intense pining. Even on a good day it was as if everything around me had been transformed. The sun never shone, the birds didn't sing, even the larch's slender branches hung down like a mourning veil over the silent, still dank woods. On those awful, aching, melancholy days, when the summer breeze wafted across the carpet, reviving mixed odours and evoking memories, I would find myself trying to catch the wind, knowing that as it travels everywhere, it *must* have touched Breeze.

Just as I knew Breeze could not have survived without our help, I was beginning to doubt my own survival without *her*, and when the postman delivered the snaps of Breeze's last days with us, I felt my heart would never be whole again.

The days which had never been long enough to fit in all our various tasks now stretched to eternity, and my sense of misery deepened. The medical profession's diagnosis of 'clinical depression' saw all of three months pass by before I finally surfaced.

As the summer of unbrowsed garden foliage faded into September, the swallows began gregariously gathering on the telegraph wires, a sure-fire indicator that the girls' sun-beds would shortly be brought inside. I felt deeply saddened at this thought, for with Sheba becoming more delicate by the day, one of the pleasures of life left to her was lying outside until long after the sun vanished, watching the silhouettes of sheep grazing around her, and soaking up the warm night air.

The only thing I was aware of as I let the girls into

the paddock on a later, cool September morning was the deafening din from flocks of swooping swallows, all vying for space on the already congested telegraph wires. Even as I looked, the chattering congregation swelled as if all compelled by some magnetic force to meet at Coldside.

It was migration time. Suddenly, the birds soared as one, twittering on the wing, then, swooping down toward the river, began following its southerly course. Soon they would be crossing continents. I continued staring after this mass exodus until not a single swallow was left in sight.

Turning to join Emma and Sheba in their mooching, we all heard the sound simultaneously. As my heart pounded, all three of us turned and fixed on the hilltop. A pair of eyes, two velvet pools exuding tenderness, were shining down on me. My breath came out in a gasp. 'B..r..e..e..z..e.'

Instant recognition flickered in Emma's eyes, and she bounded forward to greet her long-lost friend. I couldn't move, I seemed locked to the ground. When Breeze brushed up against my thigh, I felt only the lightest whisper of a touch. Apart from her initial 'peep', the greeting was strangely silent, almost ethereal, somehow.

Suddenly she was back on top of the hill flexing her neck, not its usual full three hundred and sixty degree turn, but upward and backward as though looking over her shoulder and pausing every few seconds to study my face. There was something in her steadfast look, as though a command was being directed toward me. It was with the next toss of her head that everything suddenly became crystal clear.

Her gestures were beckoning, urging me to follow. My presence had been requested. I had no need to ask the whys and the wherefores, instinctively I knew, and as I followed my heart up the hill, Emma and Sheba trotted excitedly in my footsteps.

Breeze inched her way silently towards the forest as though on tiptoe, to within a few feet of the perimeter fencing. Then, with a trusting look in her eye, she raised her right foreleg, and as she placed it down gently, I knew this was as far as she wanted me to go.

I dropped to a sitting position and held the girls close to my side, my eyes glued to the forest. I watched Breeze soar over the fence. I swear my breath stopped as small puffs of vapour rose from the bracken and two little fawns emerged from their camouflage.

As she proudly nuzzled her dainty dappled twins, bestowing upon me the ultimate compliment, I desperately wished Bernard was here with me, sharing this honour. With her head elevated, and the instantly named Dawn and Dusk at heel, they faded silently from view in a direction determined by the call of the wild, their destined pathway of life. Of all the paths and memories trodden through my mind, this one would wind forever.

21
Paradise Lost?

'The Government has asked the Forestry Commission to dispose of 250,000 acres of woodland over the next ten years. The motive for this sale of public land is to reduce the call on the public purse and make the Forestry Commission Enterprise more efficient. This ends the early evening news. Now over to the weather forecast.'

The words of the TV newscaster rang in my ears...

'Yet more loss of wildlife habitat,' Bernard stated bitterly.

'Wouldn't it be wonderful if, instead of large timber companies buying up the forests for financial gain, people like us bought them for their conservation value?'

'People like us can't afford them, Marie.'

'No, I guess not,' I answered, dispirited.

'Cheer up, pet, we may win the pools next week.'

Although Bernard made light of the situation, my mood remained disconsolate.

The following day an unexpected visitor was to chase my blues away.

Five weeks had elapsed without a glimpse of Breeze. Accepting that all contact was broken, and that never again would she grace our garden, we were to find everything is possible in Breeze's case.

Baba and Sheba were relishing the unusual warmth of a late October afternoon when, from out of the blue, she landed.

Feeling obliged to help us in our pruning of the rose bushes, she filled her tummy to bursting, then collapsed on the sun-bed. Tucking her legs under her body, assuming her sleeping position, she accompanied the girls in a nap.

My immediate thoughts were of Dawn and Dusk, left alone without a baby-sitter. Concern, then panic set in. Some dreadful fate must have befallen them.

'Don't be silly, Marie, they will be nearby. Remember how Breeze lay for periods concealed in the rushes?'

It was almost two hours, for my part feeling utterly miserable, before Bernard's head cocked to one side.

'Did you hear a distant "peep"?'

'No, did you?' I answered hopefully.

'Ssh, there it goes again.'

I didn't get a chance to strain my ears. Breeze had responded instantly to the cry, and had taken off without so much as a fond farewell.

Enjoying her occasional freedom from responsibilities, Breeze's brief visits went on for some weeks. Suddenly, though, and irrevocably, her intimate contact was broken.

On November 20th, the last call of nature came, taking Sheba softly in her sleep. For the third time we found ourselves robbed by death.

Gone was Emma's desire to delight in the sun's rays alone. The sun-bed was put away and the door closed. She didn't adjust to life without Sheba. Like us she grieved. A solitary animal pining for its own kind is a pathetic sight indeed. Emma needed canine companionship.

When first moving to Coldside, Bernard had helped

out at the local animal refuge. Of one fact he was sure, there wouldn't be any shortage of dogs desperately deserving of a loving home.

The route to the refuge led us alongside the back of Coldside Wood. Bernard took his foot off the accelerator and began coasting. Craning my neck sideways, peering into the trees, I was totally unaware of what was happening directly ahead.

The car came to a sudden standstill, jolting my head forward. Bernard switched off the engine. In full view, only metres away, a deer leapt out of the forest, landing in the crown of the road. Attempting to squeeze through a gap in the dry-stone wall, both at the same time, two immature fawns were following, oblivious of the perilous barbed wire hovering over their heads.

Only one scrambled through, bounding to its mother's side.

Directly the pitiful shrill 'peeps' pierced the air, Bernard coolly responded to the cry.

'She's entangled in the barbed wire, Marie. Please, please stay in the car and leave it to me.'

Acknowledging my limitations, being more of a hindrance than a help in a crisis, I offered no resistance and remained in the car, wringing my hands in anguish.

As I watched through the windscreen, Bernard advanced swiftly towards the struggling fawn. My heart reached out to the doe, circling in agitation, as she shielded the other fawn, shivering in fear at the nearness of a human.

'Hello, little one,' Bernard hailed her softly. 'Are you my Breeze?'

Nose twitching the air, searching his scent, she focused on Bernard. In definite recognition, her eyes drained of fear. Breeze stayed perfectly still as Bernard climbed over the fence, trusting her helpless fawn implicitly to Bernard's care.

At this point I heard a vehicle approaching from

behind. Hurling myself into the road, using my body as a barricade, I furiously flagged the car to a halt.

The astonished occupant was the ranger's wife.

'Would you please wait awhile? It's Breeze and her fawns, one's got herself caught on the barbed-wire fence, and Bernard is rescuing it.'

'It looks like he's been successful,' she replied with a warm smile.

As Bernard placed the fawn gently down on the roadside I knew her well-being was assured. She promptly fled to join her mother, and Breeze caressed her with a loving tongue, sharing with us her pleasure in her little one's release.

'Well, she appears none the worse for her misadventure. Wasn't it fortunate you chanced this way?' the lady remarked, watching Breeze, Dawn and Dusk move slowly away, disappearing into a grove of trees.

My soul answered silently, reinforcing my unshakeable belief in a providence which could never desert her.

'Apart from a lump of fur missing and a slight graze at the base of her neck, she was otherwise uninjured, thank God,' Bernard told me on our ride to the refuge, both of us intoxicated by what had occurred.

Julie, the kennel girl, introduced us to a beautiful big, bold German shepherd who had been in their care for four months. She launched herself immediately at the dividing wire-netting with all the ferocity of a lion, yet her tail waved wildly, so we knew she had chosen us. Her information sheet stated, 'Street walker. Age unknown. History unknown. Name unknown.' All we did know was that someone no longer wanted her.

Fortune has a way of smiling on us!

She didn't have the advantage of youth, being 'about' seven years old, but did have 'one hell of a pair of lungs'. A perfect companion for Emma, we agreed, as we settled her into the car and drove happily home.

It was clear from the outset that Sheba Two, as we named her, deemed herself to have landed in clover, as she never put a foot wrong from the moment she entered our home. True to character, Bossy Boots barked her protest as she neared her bed. Undaunted, Sheba Two plonked herself down with such an air of self-assurance, that Baba was taken by complete surprise, even moving her bum, allowing her companion more space.

Emma's zest for life was instantly rekindled, thanks to her new friend with the familiar name, wearing only a different coat.

We were endowed yet again with a Christmas of richness and depth.

The chimes of New Year heralded in a mildness, like spring. Having fed the multitude of garden birds, I stood awhile fascinated, watching a rook riding on the back of a sheep, partaking of his breakfast, purging the sheep's fleece of ticks.

A sound, then a branch springing jerkily, caught my attention. Just observable, screened by a holly bush, was a deer staring fixedly towards Coldside. With growing excitement I awaited the arrival of our special guest, perhaps three. For a few moments she stood, neck fully extended, seemingly searching the garden, then she turned, and slipped silently from view.

The happiness engendered by this brief encounter was as fulfilling as any visit. Although infrequently sighted, her presence remained with us throughout the mild winter.

It was a mad March hare that led us well into the woods, bringing us to a halt at the stone-wall perimeter. Directly beyond, on Eddie's land, was a rocky outcrop, and to the left a well-sown crop of luscious grass. It was to this field I turned my attention, spying three deer busily grazing. Reaching into my pocket, I extracted my binoculars and focused on the spot. A roe buck and two yearlings. Where was the doe? I studied every inch of the field, concluding she was resting in the confines of the firs.

Turning away, now head on with the outcrop, I was face to face with a perfectly composed doe. Emma, unfailing in her remembrance of Breeze, responded with a wag of her tail. Necks outstretched, they met each other's scent mid-air, then Sheba Two stuck her head over the wall. Instantly realising it was not the Sheba she knew, Breeze bounded towards her family, leaving Emma still sniffing the air.

Alerted, the fine young buck's head shot up, his six-point antlers still in velvet, rigid, his eyes searching. The moment he caught sight of me, he was back in the wood, his entourage following behind. My blessings went with Dawn and Dusk who were shortly to follow a life of independence in the predatory wide world. Breeze, under the protection of the close season, and back on home ground, would soon be seeking some secluded spot to give birth to her young. Alas, for the buck, his dodging was about to begin.

The very next day, a chance meeting with someone involved in the protection of badgers, informed me of the unbelievable. The *Estates Gazette* of March 2nd, under the heading of 'Government surplus land for sale', bore testimony to the forthcoming sale of Coldside Wood.

Our world fell apart.

What of Breeze? What of her family's future, and that of the many other creatures that had made this woodland their home? We were stunned. It just couldn't be true.

Then the large block ad appeared in the local newspaper. Coldside Wood was 'FOR SALE'. On the open market, to the highest bidder, closing date for tenders, August 21st.'

When the particulars from the selling agent reached us, stating the wood's geographical accessibility to well-established, major, small roundwood timber markets in the area, accompanied by the words 'yield class', horror struck deep into our hearts. We could not surrender to apathy. We would be failing the woodland inhabitants if we did.

We communicated frantically with several private caring organisations that had a genuine interest in acquiring areas of land with a conservation intent. We hoped they could take another valuable habitat under their umbrella, but although their compassion said 'yes', their funds said 'no'. It was solely up to us, two little individuals, to find a way to purchase Coldside Wood.

We surveyed our house contents. We could sell this and that, and exchange the car for an old banger, but even then we would be left with a shortfall of many thousands. The only means of finding the enormous sum of money to secure the woodland in such a short time was to mortgage our home.

Important decisions for us require no procrastination. We immediately contacted the building society, stressing 'time is of the essence', placing the question of how we could meet the repayments 'on ice'. What would it matter anyway, if we couldn't manage the wherewithal? To the inhabitants of Coldside Wood *their* home was priceless.

Then came a ray of hope. We learned that the Forestry Commission, having instructed the district valuer to assess the value of Coldside Wood, could offer it to us at its valuation price, not a preferential price. However, we had to have an endorsement from either of the public bodies, The Nature Conservancy Council or The Countryside Commission. Our hopes were dashed as the bureaucratic treadmill ground to a disappointing halt. 'Whilst your aims are laudable . . .'

Coldside Wood was adjudged too insignificant a woodland to warrant their patronage.

BREEZE: WAIF OF THE WILD

The news soon got around that we were mortgaging our home to enable us to bid for the wood, solely for its conservation. The local press hot-footed it up to Coldside. Border TV's newscaster and a camera crew followed closely on their heels, and the following day our commitment to preserve the habitat which provided shelter and food source for a variety of animal species was broadcast news.

Within a matter of weeks national newspapers had picked up on our mission, and the 'good luck' letters started pouring in. One such letter had an enclosure of £2, and apologised for it being such a minuscule amount; it was signed 'Old Age Pensioner', and it brought tears of humility to our eyes. Another came from a lady in Suffolk who had tried raising funds to save a local wood. She had, alas, failed, and ninety-eight houses now stood where wildlife once thrived. She sent £10, along with words of great encouragement: 'Better to have tried and lost, than not try at all.'

Many letters arrived with only the postmark giving a clue as to the sender. One such, containing a book of ten new first-class postage stamps, folded in a small scrap of paper, with the words 'Good luck, from a ninety-year-old forest friend', touched our hearts. Another morning, postman Bill called with a carefully wrapped package. It wasn't addressed to us, simply 'The Coldside Wood Animals'. When we opened it, brightly coloured stamps of all denominations from various countries of origin spilled out. The note enclosed said, 'My name is Jonathon. I am twelve years old, I love all little living creatures, and would like very much for you to save their home.' That Jonathon was willing and happy to part with his treasure trove showed true love indeed.

There were concerned celebrities too who rallied round, sending letters of encouragement and personal items for us to sell. Bryan Robson (a signed England shirt), Henry Cooper (a signed tie), Christopher Timothy (a signed shirt, worn in the series *All Creatures Great and Small*), Rowan Atkinson, Jilly Cooper, Richard Briers, David Suchet, Alexei Sayle, Katie Boyle, Dawn French, Gregor Fisher, Russ Abbot, Esther Rantzen, Melvyn Bragg, Phyllis Logan, Richard Adams, Bob Holness, Ruth Madoc, Matthew Kelly, Stephen Fry, Ronnie Corbett...

There were many others also whom one doesn't readily acknowledge as wildlife campaigners. Because they don't openly 'fly the flag' in the name of conservation, does not mean they are without awareness and generosity of spirit. Catherine Cookson, the novelist, was the very first cherished donor supporting our bid to safeguard this valuable wildlife habitat. Her continuing interest in the wildlife at Coldside never wanes despite her increasing health problems.

Valued friendships were forged, many only in the written word, but all worth much more than their weight in gold.

As prospective buyers armed with clip-boards and tape measures trampled round the wood, we grew more anxious by the day. Although the wheels were in motion within the building society, they were turning interminably slowly, and we found ourselves in a race against time to meet the closing date for offers. With our treasured possessions sold, we had only two days' grace before the precious news arrived. Our mortgage application had been successful. Our collateral was secured.

Dilemma! How much to bid? A pound too much was a waste of badly needed money. A penny too little meant we had lost the woodland *forever*.

Having placed our agreed offer in a sealed envelope marked 'Coldside Wood', as stipulated, Bernard took it by hand to the selling agents on the appointed closing date, naïvely believing the following morning would bring a call informing us of their decision.

The next nail-biting day came and went, with us sitting by the phone. Then a sleepless night. At nine o'clock the following morning, cracking under the strain, Bernard telephoned the agents.

'I'm sorry, Mr Kelly. We can't divulge any information, this is not the way we operate. We forward all offers to the Forestry Commission and await their instructions. You will hear in due course.'

Nerve-racking days stretched into two weeks of torment.

Convinced we had attempted the impossible, and lost the woods to the men with the chain-saws, when the letter finally came neither of us wanted to read it.

Bernard's victorious cry as he waved the letter aloft said it all. Our eyes spilled tears of joy as our hearts overflowed with gratitude. Now the trees and wildlife inhabitants were permanently safe from exploitation.

22
Spring 1995

Many changes have taken place, beginning that September, thanks to the multitude of caring people who supported our aim. A large pond has now been created, many hardwoods, shrubs and wild flowers have been planted, and owl, bat and bird boxes erected. What better way to spend our donors' gifts than where it matters most, on the woodland, and for the benefit of its wildlife inhabitants?

As dawn breaks this springtime over our magical forest, where kingcups shine golden in the early morning sun, I think back to the previous November when we last sighted Breeze. She was running her nose up and down the stem of a newly planted willow on the far side of the pond. Directly her watchful eye detected humans, she gently piloted that year's set of twins into the shadow of the pines and was gone, quietly on her way.

The links between Breeze and the girls who had been so much a part of her life as well as ours, have now been finally severed. The shadow of fate finally fell, and Emma, who had lived the breadth as well as the length of her mortal life span, was not able to share in this springtime.

BREEZE: WAIF OF THE WILD

As my fingers fly through my sewing, nineteen to the dozen, and Bernard juggles the bills in an effort to meet this month's repayment, my thoughts return to the first time I was enraptured by this forest. What would have happened to Coldside Wood if we had never moved here? Could Breeze's story have ever been told?

I firmly believe in Fate, for it has had a great influence on us... Life is full of undreamed-of surprises. We are the lucky ones, for we felt the wind blow.

More Fascinating Biography from Headline

A Greater Love
Charles & Camilla

Christopher Wilson

– THE ENDURING LOVE AFFAIR THAT ROCKED THE ROYAL FAMILY

In July 1994, Prince Charles finally admitted in a television interview with Jonathan Dimbleby that he had committed adultery with Camilla Parker Bowles. His admission shocked the nation; now, in *A Greater Love*, Christopher Wilson charts the full story of the love affair between Charles and Camilla from its early days over twenty years ago.

How could an apparently plain and unintelligent woman have captivated Prince Charles so completely for so long? How did the affair survive both Camilla's marriage to Andrew Parker Bowles, and Charles' marriage to Diana? Was it responsible for the break-up of Charles and Diana's marriage?

This sensationally frank book reveals for the first time the true character of Camilla Parker Bowles, and describes in detail the complex nature of Charles and Camilla's ill-starred romance – a romance which has gravely damaged the Royal Family.

NON-FICTION / BIOGRAPHY 0 7472 4676 9

A selection of non-fiction from Headline

THE DRACULA SYNDROME	Richard Monaco & William Burt	£5.99 ☐
PROCLAIMED IN BLOOD	Hugh Miller	£5.99 ☐
MURDER BOOK OF DAYS	Brian Lane	£6.99 ☐
THE MURDER YEARBOOK 1995	Brian Lane	£5.99 ☐
THE PLAYFAIR CRICKET ANNUAL	Bill Findall	£3.99 ☐
KEITH: TILL I ROLL OVER DEAD	Stanley Booth	£5.99 ☐
THE JACK THE RIPPER A–Z	Paul Begg, Martin Fido & Keith Skinner	£7.99 ☐
THE *DAILY EXPRESS* HOW TO WIN ON THE HORSES	Danny Hall	£5.99 ☐
AT HOME WITH FRED	Rupert Fawcett	£5.99 ☐
GRAPEVINE; THE COMPLETE WINEBUYER'S HANDBOOK	Anthony Rose & Tim Atkin	£6.99 ☐
THE LEX FAMILY WELCOME GUIDE TO HOTELS, PUBS AND RESTAURANTS	Jill Foster & Malcolm Hamer	£7.99 ☐

All Headline books are available at your local bookshop or newsagent, or can be ordered direct from the publisher. Just tick the titles you want and fill in the form below. Prices and availability subject to change without notice.

Headline Book Publishing, Cash Sales Department, Bookpoint, 39 Milton Park, Abingdon, OXON, OX14 4TD, UK. If you have a credit card you may order by telephone – 01235 400400.

Please enclose a cheque or postal order made payable to Bookpoint Ltd to the value of the cover price and allow the following for postage and packing:

UK & BFPO: £1.00 for the first book, 50p for the second book and 30p for each additional book ordered up to a maximum charge of £3.00.
OVERSEAS & EIRE: £2.00 for the first book, £1.00 for the second book and 50p for each additional book.

Name ....

Address ....

....

....

If you would prefer to pay by credit card, please complete:
Please debit my Visa/Access/Diner's Card/American Express (delete as applicable) card no:

Signature .... Expiry Date ....